VHDL: A Logic Synthesis Approach

VHDL:
A Logic Synthesis Approach

David Naylor
Aspen Consultancy Ltd,
London, UK

and

Simon Jones
Professor of Electronic Systems Design,
Loughborough University,
Loughborough, UK

CHAPMAN & HALL

London · Weinheim · New York · Tokyo · Melbourne · Madras

Published by Chapman & Hall, 2–6 Boundary Row, London SE1 8HN, UK

Chapman & Hall, 2–6 Boundary Row, London SE1 8HN, UK

Chapman & Hall GmbH, Pappelallee 3, 69469 Weinheim, Germany

Chapman & Hall USA, 115 Fifth Avenue, New York, NY 10003, USA

Chapman & Hall Japan, ITP-Japan, Kyowa Building, 3F, 2-2-1 Hirakawacho, Chiyoda-ku, Tokyo 102, Japan

Chapman & Hall Australia, 102 Dodds Street, South Melbourne, Victoria 3205, Australia

Chapman & Hall India, R. Seshadri, 32 Second Main Road, CIT East, Madras 600 035, India

First edition 1997

© 1997 David Naylor and Simon Jones

Printed in Great Britain by Cambridge University Press

ISBN 0 412 61650 5

A catalogue record for this book is available from the British Library

∞ Printed on permanent acid-free text paper, manufactured in accordance with ANSI/NISO Z39.48-1992 (Permanence of Paper).

Contents

To

Archontia Mexi
and
Lisa Naylor

Preface

Another book about VHDL. Evidence either of its importance or the ease with which the subject can be written about, depending on your sensibilities. We fall in the former camp now; previously we would have seen merit (in the impartial manner so fondly adhered to by some academics) in both camps. The difference between possibility and practicality, which engineers are perhaps uniquely placed to judge, has been brought home to us by the precision and revisions necessary to transmit not only our experience in using VHDL, but also the pedagogic constructs necessary. In short it was easier to learn how to program in VHDL than to learn how to teach it.

Design is an experiential subject. You learn it by doing it. This book is an aide-memoire, a travelling companion for those who wish to use VHDL. It will not solve your problems for you, but it will help you arrive at your own solutions in a faster and , we hope, elegant manner.

All the VHDL code in this book is accessible via the World-Wide Web. The URL for the code is http://info.lboro.ac.uk/departments/el/research/sys/index.html. At that page there can be found a hypertext link entitled 'VHDL – a logic synthesis approach – VHDL code'. Clicking in this entry will take you to a number of pages, one per chapter. Any email comments, queries or feedback can be made to S.R.Jones@Lboro.ac.uk.

Like film making, producing a book is most definitely a joint activity and we would like to thank many people who contributed to its design. Christine Armstrong, Mark Gooch, Julian Yeandel, David Thulborn, Morten Kjelsoe, Mohammed Saeed, Karl Sammut, Robbie Temple and Ryan Lim of the Electronic Systems Design Group for their feedback on the text. Bill Lockley, Lynne Bennett, Hannelore Nunn, Rob Brackley, Elena Trichina, Richard Evans and Heiko Schroder for the moral support during the time that the book was being written. Mark Turner at Veribest for access to the synthesis software and libraries. In particular we appreciate greatly the help of Jon Talbot for his technical support. David Naylor would like to thank his wife Lisa for agreeing to marry him, even though this book was only half complete – a brave woman indeed! He would also like to thank everybody who asked 'How is the book going?' at every occasion. Without their persistence his life would have been far less stressful! David would like to thank Simon Jones for his support and assistance, which have made this book possible. Finally, Simon Jones would like to thank David Naylor for making this book far easier to write than it would otherwise have been.

Like food, wine, travel, poetry and other fine things in life, good design is best experienced not described. We hope that you enjoy your experiences.

Loughborough, UK
May 1997

1

Introduction

1.1 PREAMBLE

This is a book about the language VHDL. VHDL stands for VHSIC hardware description language. VHSIC is itself an abbreviation for very high-speed integrated circuit. VHDL is a language that can be used to describe the behaviour, structure and implementation of electronic systems. As such, VHDL is a computer-aided design (CAD) tool. The use of design tools for the construction of electronics has a very long history (at least by the timescales of the information technology industry). The role of these tools has expanded as the phenomenal growth in semiconductor technology has changed the primary challenge faced by engineers from the mastery of detail to the management of complexity. It is probably fair to say that electronic systems designers today are limited more by their capacity to handle the complexity of their designs than by the ability of the technology to support it. VHDL is a design tool that aims to manage this complexity (by allowing a high-level design description to be supported) and, through the use of logic synthesis tools, to generate automatically a hardware structure that meets the specification. As such, we believe that VHDL will become the predominant design method for the construction of advanced electronic systems. Digital electronic engineers must in the future have both familiarity and facility with VHDL in order to practise their profession. This book aims to meet this need.

There are many books on VHDL; we have many of them in our personal and team libraries. However, many of these books focus on VHDL as a specification and simulation language. Unfortunately, as practising research engineers we found that when learning VHDL these books did not exactly meet our needs: not because they are incorrect or misleading, but because they focus on the use of VHDL for simulation. Although, naturally, we used VHDL for this purpose as well, we were designing, to a very tight timescale, a 500+ processor machine for real-time image processing and were using logic synthesis tools to compile our hardware from a high-level description of its behaviour. We realized that the capabilities of logic synthesis tools, impressive though they are, restricted both the range of VHDL constructs that could be compiled and the style of VHDL specifications that they were written in. Through a period of trial and error, we taught ourselves these rules and accumulated some experience in writing synthesizeable VHDL. We decided to present our experiences in a book aimed at teaching VHDL for what we felt would be its main application in the next decade, that is as an input into logic synthesis CAD tools. As such, this book focuses on a subset of VHDL and on styles of coding that are best suited to synthesis tools.

At the end of this book, the reader should be able to successfully write and implement quite complex designs using VHDL; understand how, when and where to apply particular approaches; and, equally important, how, when and where not to do so. We are enthusiasts and great believers in the importance of VHDL. We hope this book communicates this enthusiasm; please enjoy it.

1.2 COMPUTER-AIDED ELECTRONIC SYSTEMS DESIGN

Computers have been used in the design of electronic systems for over 30 years. One of the earliest (and still a widely used class of tools) is the analogue circuit simulator, best exemplified by SPICE – the simulation program with integrated circuit emphasis. SPICE was developed in the mid to late 1960s at the University of California, still today a major centre for the research and development of CAD tools. SPICE enabled users to model a wide range of electronic devices through the description of a circuit as a netlist. A netlist is simply a topological description of the way in which the various circuit elements are connected together. Each element is described as a basic model (e.g. a resistor, capacitor or transistor), which details the electrical parameters of that component, together with the particular characteristics of that circuit element (e.g. the channel width and length of a MOSFET). The SPICE description was originally assembled as a deck of punched cards and then simulated. The results of this simulation, usually a string of floating point values, were read or interpreted by the user.

As computers became more plentiful and more powerful, design entry tools and design visualization tools became available. Text editors and time-sharing systems enabled a user to edit the SPICE description as a text file and interactively simulate the circuit. The introduction of graphic display terminals and output devices such as low-cost plotters enabled the simulation results to be displayed in a visual and hence easier to comprehend manner. Even today, nearly 30 years on, SPICE and variants of SPICE are still widely used in the simulation process.

Following on from this, the growth in digital design demanded simulators that could cope with the increasingly complex circuits that engineers were producing. Analogue simulation of large circuits was and still is computationally demanding. Digital designers needed assurance that the logic of the systems was operating correctly. By the mid to late 1970s digital simulators were becoming available to fill this role.

Of course, simulation is only one part of the design process. Design entry is also important. Designs were either described algebraically, as in Boolean equations, or were drawn manually using draughting techniques. By the early 1980s the price/performance of workstations and personal computers had improved sufficiently to enable draughting programs to provide automatic schematic capture and, for integrated circuit design, layout tools enabled the drawing of physical structure of the circuit to be accomplished on-line.

However, a key problem still persisted at this time: how to ensure an isomorphism between the different design descriptions or, less prosaically, how to ensure that the circuit schematic drawn was the same as the circuit simulated. By the mid-1980s software packages that automatically moved from schematic design through to translating the design into a language that a simulation package could accept were available. Working from the bottom up, circuit extraction software became available, which took an integrated circuit layout and extracted the actual circuit topology, together with detailed electrical information (e.g. parasitic resistances and capacitances), enabling both the verification of layout and more accurate analogue simulation. This same period saw the rise of major companies that provided the toolset for electronic design automation (EDA). Many of them, such as MENTOR and INTERGRAPH, are now substantial corporations.

Simulation is, of course, only one side of the problem. Another problem is entering, modifying, checking and verifying the actual design. The introduction of schematic capture did much to address this problem and is still widely used today. However, for the last 20 years or so the number of devices that can be integrated on a single integrated circuit has been doubling every 2–3 years. This technological improvement, combined with the user-led demand for more complex systems, for applications ranging from washing machines, through mobile phones, to personal and portable computers, has meant that the complexity of the design tasks has grown at a similarly phenomenal rate.

The drawback with schematic-based approaches arises out of their ability to represent structure. For smaller designs, the function or behaviour can easily be comprehended from the structure. However, as the designs get bigger, the number of components, the range of the functions and the richness of interconnection between them makes it increasingly harder to extract function from their structure. Such techniques as hierarchical schematics help of course, but can only postpone the problem when complexity is doubling every 2–3 years. To manage systems with complexities of the order of hundreds of thousands to millions of components, another approach was sought. Perhaps the only other discipline in modern life that has to handle highly complex systems is software engineering, and the language-based approach of that domain appeared to many researchers to be an attractive way forward.

1.3 HARDWARE DESCRIPTION LANGUAGES

A hardware description language (HDL) is a means of describing some or many attributes of an electronic system in a text-based language. The majority of contemporary HDLs appear to be remarkably similar in syntax and constructs to high-level programming languages such as ADA, C++ and PASCAL. However, the design of an electronic system is conceptually different to the design of a software package. These differences include the following.

1 An electronic system is a concurrent parallel system. All circuits in an electronic assembly are continuously operating. In contrast, software systems incorporate the notion of program flow, in which the sequence of instructions executed is explicitly stated through features such as loops and conditional statements.
2 A software package is almost exclusively concerned with the design of the behaviour of the system. Although factors such as program size and speed are accounted for in some packages, the situation in electronic systems design is different. For example, the engineer is concerned not only with behaviour, but also with the structure of the system (perhaps to minimize package count or clock period), the location of functions (perhaps to ease system partitioning problems) and the underlying technology chosen (perhaps to minimize power consumption)
3 In electronic systems design the opportunity for real-life system testing is greatly restricted. This is in contrast to software engineering, in which multiple recompilations can be undertaken with little extra cost. In a project involving a complex integrated circuit design there may be only one opportunity to go to silicon and evaluate the real-life performance. For this reason, there is a much greater need for the language and the support tools to assist in detailed design simulation and evaluation

As a result of these differences, the early use of register-transfer languages and similar notations had a limited impact on electronic systems design. However, experience gained in the design and application of these early languages, allied to the increasing demand for language-based design as a means of managing complexity, meant that by the late 1970s and early 1980s languages such as HILO and, within the UK, ELLA were being increasingly widely used as design entry and simulation tools. From the United States, the Department of Defence (DoD) push into advanced microelectronics as a means of enhancing the capabilities of weaponry led to the formation of a grouping with the goal of developing a language for the documentation of complex integrated circuit designs. In many ways, this approach reflected the perceived success of the ADA language, which itself had been developed as a DoD-inspired common language for defence software.

Initially, VHDL was designed as a language for documenting designs. However, the language specification being in the public domain, allied to the strong research push in microelectronics design in both Europe and the USA, led to the language being used as a simulation tool by research groups and industry. Indeed, today a wide range of designs are modelled in VHDL. Moreover, its use has expanded beyond electronics. We have even encountered the language being used as a tool to model human factors design problems! However, the size of the electronics market has meant that the primary use of the language today is as a tool of electronic systems design. At the time of writing VHDL is an IEEE standard. The current standard 1076 defined in 1993 is the version of the language this book uses.

1.4 VHDL

VHDL is a hardware description language. Part of its success can be attributed to the fact that the designers of the language recognized the particular requirements of electronic systems design. A further part can be attributed to the designers incorporating recent developments in programming language design, including object-oriented techniques and strong typing. Some of the differences between VHDL and a sequential programming language such as C/C++ are as follows.

1 VHDL is a parallel language, i.e. there is no implicit default sequential control flow as in software engineering. Each and every VHDL statement can be executed concurrently, unless otherwise specified. Constructs exist in VHDL for explicit sequential steps, but they are precisely that – explicit language constructs.
2 VHDL is a strongly typed language. This is in contrast to many C variants, in which operations on mismatched data types can be supported. In VHDL, typing is strongly enforced; where variables of different types need to be assigned, explicit casting needs to be performed. This is very much like the situation in a real electronic system, in which, for example, an non-tristate output pin should not be connected to a bus line.
3 VHDL allows the use of explicit time delays. It is possible to say that a statement is executed after a certain time delay. This is in sharp contrast to programming languages in which there is a concept of sequence but not delay. Specific time delay procedures are often supplied as part of a programming language support suite, but they are not part of the language themselves. VHDL constructs explicitly model the time delay inherent in all electronic circuits. At this point it is perhaps pertinent to

mention that this feature of VHDL is one of the most fundamental differences between the use of VHDL as a simulation language and a synthesis language. No synthesizer is able to construct a circuit with a fixed time delay. The actual time delay will depend on the technology, which may not be known at the time of synthesis, or the temperature and process variations, which vary continuously.

VHDL also has many similar attributes to object-oriented programming languages such as ADA and C++. These similarities include the following.

1 The object-oriented emphasis on the separation of the interface of an object from its operation is mirrored closely in VHDL. In this case the entity provides the interface specification to the circuit and the architecture (implementation) of the circuit provides the functionality. Moreover, the same entity can have multiple architectural representations, thus providing a range of descriptions for the same component. Typically, this is of use in top-down design, in which a behavioural description can be modelled first and the design incrementally refined towards a more detailed structural description. Where logic synthesis tools are involved it is possible to have multiple descriptions of the same circuit and it is easy to experiment with the synthesis tool to find out which instantiation of the circuit function is the most suitable.

2 VHDL is a strongly typed language and hence mirrors the explicit typing in ADA. Strong typing is considered by software practitioners to be an important attribute. This feature is also exhibited in VHDL. Moreover, the fact that VHDL is a language for electronic design means that the physical characteristics of signals need to be clearly indicated, as the type of circuit element (e.g. input, output or input–output) has a strong effect on structure and performance.

3 Operator overloading is the capacity found in many object-oriented languages to specify the effect of operators. This can be used to define the behaviour of new operators and to redefine existing operators more closely to the requirements of the system being designed and the technologies being used.

1.5 SYNTHESIS OR SIMULATION

1.5.1 Simulation

VHDL originated as a design documentation language. Quite early on, its role as a simulation tool was widely appreciated. When designing systems using commercially available components, it is possible either to purchase a library that provides a VHDL simulation model of the components used (this can sometimes be provided by the vendor) or to write one's own library based on data sheet information. Once the VHDL models are available the design process is fairly standard. A design description is constructed. This design description could be a schematic-based design in which each element in the top-level schematic is a VHDL program, or more usually a VHDL program that structurally connects these elements together. Where more complex functions are involved, the design may start out as a behavioural description, which after verification can be re-engineered using available components. This design is then compiled and simulated. When the design is functionally correct, it can be passed through to the floorplanning, routing and placing tools. The delay information extracted from this placement can then be fed back into the VHDL model for post-layout verification.

In this mode VHDL is used exactly in the same manner as any circuit or logic simulation package. A design description is compiled and simulated and the results evaluated. In this mode VHDL offers the same productivity gains as can be expected from any CAD suite. Things are made easier owing to the possible wider availability of VHDL models and the wide use of VHDL, enabling design teams to reuse other designs, but essentially the process remains the same and the benefits are those of standardization and wide use.

1.5.2 Synthesis

In this route VHDL is used as the specification and synthesis language, but the process of transformation into hardware is substantially more automated. Specifically, the process of high-level simulation is maintained by the transformation from this level of description into hardware elements being automated. Generally, logic synthesis software produces output in the form of a netlist. Hence, such software's output is often targeted to FPGA or ASIC technologies.

With the synthesis route, the process of mapping into particular integrated circuit technologies is automated. The synthesis process therefore both increases productivity (through automation) and reduces flexibility (by restricting the choice of technology).

Furthermore, the analysis of the system design at an early stage is more complex. When choosing the simulation route, the number of integrated circuits could be known at a relatively early stage if, for example, each architecture in the library was a single integrated circuit. With the synthesis route, it is only after compilation and optimization that the hardware size and speed can be accurately known. Although experience can help in judging the eventual size of a piece of code, small changes in the code can produce remarkable changes in hardware size. Several chapters in this book address the issue of suitable design styles for different types of circuits.

1.6 LOGIC SYNTHESIS

So far this chapter has mentioned logic synthesis without precisely defining it. The general view of logic synthesis software is of a CAD tool that can take a VHDL program and produce a netlist to realize exactly that function. As such, synthesis tools are probably the most advanced and complex design aid available today. Like almost everything else in the electronics industry this field is evolving rapidly, and the capabilities and constraints of logic synthesis software are changing daily. Nonetheless, some salient facts exist. Firstly, not all VHDL can be synthesized. There are a number of reasons for this.

1 It is possible to specify behaviour that is impossible to implement. For example, two numbers are added after exactly 3.2 ns. It is clearly impossible to meet these conditions, as the time delay will depend on technology, circuit design, what numbers are being added, the number of bits involved, the temperature and power supply, to name a few parameters.

2 It is possible to write VHDL code that the synthesizer cannot cope with. For example, relatively few synthesis tools can support division operations. Furthermore, some VHDL design styles are impossible for the synthesizer to analyse successfully. As time progresses, research into these issues means that the capabilities of the software are continuously improving, but nonetheless substantial limitations exist. It is a major aim of this book to communicate techniques that enable the writing of VHDL code that can readily be processed by logic synthesis software.

1.7 STRUCTURE OF THIS BOOK

Following this the first chapter, the second overviews logic synthesis, explains the design cycle and reviews the main types of synthesis tools available today. Design optimization and simulation techniques are introduced. Chapter 3 introduces the principles of VHDL by focusing on the syntax of the language and the way in which different constructs are used together to solve design issues. Chapters 4, 5, 6 and 7 address combinational, sequential, arithmetic and control circuit design respectively. The eighth chapter brings together the different strands through a major design example. Chapter 9 concludes with a summary of the main points and highlights future trends and issues in the automated synthesis of electronic systems.

2

Logic synthesis

2.1 THE DESIGN CYCLE

The traditional way to design electronic circuits is as follows: firstly, specification of the behaviour of the systems; secondly simulation of the behaviour; thirdly, transformation of this behavioural description into a dataflow; fourthly, structural description, verifying along the way that each decomposition stage maintains the exact behaviour as detailed in the behavioural specification. The final stage is then to transfer the structural description

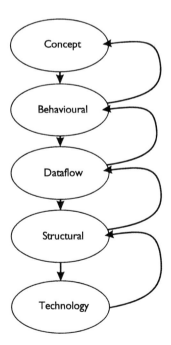

Figure 2.1 Top-down design cycle.

into a technology design. Figure 2.1 shows the design cycle in a diagrammatic form.

The arrows linking each stage to the previous one show that at each transformation there is the possibility, indeed the likelihood, of errors being introduced as a result of the manual decomposition. Of course, conventional CAD tools can do much to reduce and detect the errors introduced at each stage, but the manually instigated translation process inherently leaves open the possibility of errors being introduced. The automated nature of synthesis reduces this possibility. Now, of course, this does not mean that going from a concept to a design is a one-step process in VHDL. Quite the contrary, as the design behaviour itself might be inappropriate or the estimates or actual speed and size figures might proscribe such a design description, requiring rework at the design description level, but this is a different issue from design errors. With logic synthesis we can go with a high level of assurance from a description to a design. This factor is the strength and the source of the productivity gains in using VHDL-based logic synthesis.

Synthesis is often defined as the process of translating from a behavioural to a structural description. For VHDL this description needs to be filled out a little since VHDL allows the description of a system from many different viewpoints, for example, behavioural, dataflow or structural, or indeed a mixture of these three approaches. Hence, a VHDL synthesis system must translate for any or all of these descriptions into a technology net-list. In practice, this technology can be a generic format (i.e. not linked to any specific technology but using a core set of Boolean and storage primi-

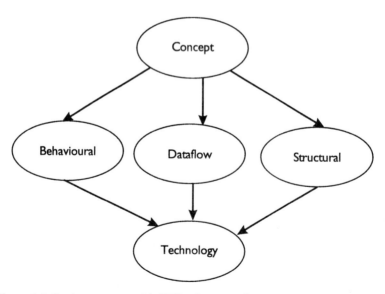

Figure 2.2 Design process with VHDL logic synthesis software.

tives) or an actual technology. Figure 2.2 shows the design process with synthesis software.

The absence of reverse arrows indicates that the design process does not need to iterate because of faulty transformations from one description to another. In practice, getting the design wrong, having too big or too slow a design, will result in further modifications of the descriptions.

Given that VHDL synthesizers can accept and process a range of design descriptions and result in a technology net-list, one might wonder why the dataflow and structural approaches need to be used, as the behavioural description gives the greatest design leverage. If one can write 'c <= a*b;' and get that transformed into a design, why should one bother to detail a structural description of a Baugh–Wooley multiplier, for example? The answer lies in the conflicting issues of design productivity, speed and size. The behavioural description gives provides the greatest design productivity, but, as many of the chapters in this book show, the limitations of contemporary synthesis algorithms allied to the use of general templates to define the format for particular functions means that such an implementation can be too large or too slow for the application. When this is the case then dataflow or structural descriptions can be used to create more optimal structures at the expense of increased design time.

Of course, a design is rarely produced from a completely blank piece of paper, with a team of designers all using the same design systems. In practice, the design team are likely to use components or designs produced using other methodologies or in different formats. Usually these are input into a synthesizer as structural or dataflow VHDL, as a result of automatic conversion software, e.g. packages that take a truth table and produce dataflow-format VHDL. Consequently, a synthesis design tool must be able to produce a design from a range of formats

2.2 TYPES OF SYNTHESIZERS

So far synthesizers have been outlined only as a tool for transformation from one level of description to a net-list of some form. In practice, it is clear that a number of different types of synthesis tools exist.

2.2.1 Architectural synthesis tools

These are the most common and widely used synthesis tools in the mid-1990s. It is this type of tool (the Intergraph tool 'Synovation') that this book uses for the design examples. Many tools of this type available from most CAD vendors. Architectural synthesis tools enable the synthesis of behavioural, dataflow and structural VHDL into a general model comprising random logic, flip flop-based storage, arithmetic circuits and control logic.

They usually synthesize into a system with limited use of bus structures. They often have limitations in their ability to synthesize largely memory-based structures and data-path structures, although the precise limitations vary from tool to tool.

2.2.2 Datapath synthesis tools

These tools are typically used for synthesizing systems into bus-based structures in which ALU and register-file-like structures are attached to a common bus. These tools are often considerably more restrictive than the more general-purpose architectural synthesizers, but for many applications that map well to a bus structure they can be much more efficient. Often a suite of CAD tools will provide both an architectural synthesizer and data-path synthesis tools.

2.2.3 Test synthesis tools

The testing of complex electronic systems is a challenging problem. As device complexity increases and more highly integrated packaging technologies such as multichip modules become widely used, the testing of an electronic assembly is becoming as important a topic as design. Furthermore, the increasingly common use of application-specific integrated circuits (ASICs), with a short production run, means that the cost of test program development cannot be amortized over a long production run. Consequently, there has been strong pressure on CAD tool vendors to provide software that can ease the testing of circuits. Within the domain of synthesis tools, test synthesis is an increasingly important technique. Test synthesis is usually based around the IEEE standard of scan-paths.

Scan-based testing is a means of improving the testability of integrated circuits by providing a special test mode of operation. In test mode some or all of the flip flops in a design are connected together to form a long shift register (scan-chain). This scan-chain can be used to read out the instantaneous values of the flip flops, which are then verified against the previously determined correct values. In addition, the scan-chain can be used to load in particular values to test out the internal logic. The use of scan-chains increases both the observability of a system (by reading out internal values) and its controllability (by setting values deep inside the system logic to a known state) and hence results in shorter test programs, which are quicker to apply and provide increased fault coverage. Test synthesis software takes a design and automatically inserts scan flip flops or converts existing scan flip flops to improve testability.

The use of test synthesis software lies outside the scope of this book, but we eagerly anticipate the arrival on the market of books devoted to this topic.

2.3 DESIGN TESTING AND VERIFICATION

Synthesis does not do away with the need to simulate a system. Simulation is necessary at many stages in the design process to verify correctness at the level of abstraction currently being developed. Within the digital electronics domain there are three main types of simulation: logic, timing and fault. The following subsections review the role of these modes of simulation within the synthesis design process.

2.3.1 Logic simulation

This is the most common form of simulation. It estimates the logical operation of the system. However, it neglects a number of important points. Firstly, it assumes that all circuits operate with either zero delay or unit delay (depending on the simulator). This, clearly, is inaccurate. Furthermore, it does not address the issues of delay related to fan-out- or fan-in. Sometimes these simulators assume that combinational logic is evaluated in a single timestep independent of its complexity. Nonetheless, despite these limitations, the logic simulator is almost always used as the first stage in a design, to verify correctness.

2.3.2 Timing simulation

Clearly, in practice, a digital circuit has delays. Moreover, these delays vary according to fan-in and fan-out (gate loading), the actual function itself and whether the change in value is high to low or low to high (fall and rise times). Further issues that need to be addressed are set-up and hold times for latches and worst- and best-case timings, which can vary considerably owing to effects such as process variations and temperature.

The purpose of a timing simulator is to check whether the logical operation of the circuit is still correct after these factors are taken into account. Timing simulation occurs after the technology net-list has been generated, as it is only after this stage that the timing information can be known. It is sometimes worthwhile running a timing simulation if a generic library is used as certain gross timing errors can be picked up at an early stage (for example when the decision to commit to a particular technology has not yet been made). If this is done it is still necessary to run a timing simulator after the technology has been chosen.

Finally, a timing simulation needs to be performed when the technology net-list has been placed and routed and the timing information is back-annotated to the design to include wiring delays.

2.3.3 Fault simulation

Fault simulators are a means of determining the behaviour of circuits when a fault occurs. They can be used to determine which parts of the circuit are difficult to test and also how many possible faults are covered by the proposed test set. As such, fault simulators assume a fault model. This fault model can vary in complexity according to the simulator and the technology used. Typically stuck-at (in which a circuit node is permanently held at one of two possible logic values) stuck-open and stuck-closed failures are simulated.

Fault simulation occurs usually after most of the timing simulation is completed. At this stage test vectors can be generated and evaluated. Where certain parts of the circuit are difficult to test it may be necessary to redesign the circuit or reapply the test synthesis tools if they are used.

2.4 DESIGN OPTIMIZATION TECHNIQUES

All logic synthesis software provides the capability to optimize a circuit for area and speed. As device technology improves it is likely that optimization for power consumption will also be necessary. Few tools offer power optimization in the mid-1990s (beyond the optimization of power provided by reducing the number of gates). After these two types of optimization a further design-based optimization is often provided called resource sharing. Finally, the way in which the VHDL code is written can have an important impact on optimization – some constructs synthesize into considerably more efficient structures than others. All optimization techniques except the last are compute intensive, requiring workstations with substantial memory capacity and processing speed. Some contemporary techniques are heuristic, and it may well be necessary to run the optimization software a number of times. Furthermore, a multiple optimization approach is often useful. Individual small circuits are optimized intensively. Intensive optimization refers to the number of separate attempts that the optimizer makes at achieving a reduction. As many optimization techniques use Monte Carlo/heuristic algorithms, increasing the number of trials will increase the chance of a better result. When all the individual components are optimized the complete circuit is then reoptimized but less intensively. This is performed because intensive optimization on a large circuit can be computationally costly. Usually speed optimization is applied after area optimization.

The following sections outline each of these four optimization techniques.

2.4.1 Area optimization

Area optimization is a well-studied field in digital electronics. Early techniques addressed simple matrix-based manipulation for Boolean algebra. Later developments incorporated optimization based on don't-care states. At the time of writing a whole host of deterministic and non-deterministic methods are available: general purpose or targeted at particular structures (e.g. programmable logic arrays). It is outside the scope of this book to address the methods, but most synthesis tools contain complex and powerful optimization methods.

Boolean logic is fairly easy to optimize, and most synthesizers perform close to optimal. Knowledge of the underlying technology strongly affects the optimization (e.g. whether or not four-input gates are available). State-machines are equally well optimized.

Where area optimization is more difficult is when particular device technologies are used. For example, the XILINX FPGA operates a number of CLBs. Each CLB can compute up to five-input Boolean expressions. If the expression involves six inputs then a second CLB is needed. If four inputs are needed, then there is unlikely to be any area saving. Here the relationship between gate count and area is non-linear and thus substantially harder to calculate. Special-purpose optimizers may need to be used for these cases.

2.4.2 Speed optimization

Some measure of speed optimization is usually achieved through area optimization, as the area optimization removes redundant circuitry and can shorten the critical path of the circuit. However, further speed optimization can also be performed.

As with area optimization, there are a wide range of techniques that are available and applied. A common approach is to identify the critical path and to work on reducing that. A simple optimization might be to replace each component on the critical path with a higher rated (i.e. faster but more power-intensive equivalent) but functionally equivalent component from the library. Other optimization techniques might be to strip out strings of inverters from a design or where, for example, AND gates are slower than OR gates to attempt to replace all ANDs with ORs.

Speed optimization is usually applied after area optimization has taken place. Speed optimization will improve the performance but only to a certain extent. If after speed optimization the circuit still does not meet the design specification, then the designer must go back and restructure the VHDL. Often this approach involves replacing a behavioural specification with a dataflow or structural one. A good example would be replacing a behavioural description of an add operation by a structural equivalent that implemented a fast carry-propagate mechanism.

2.4.3 Resource sharing

In VHDL source code there may be substantial opportunity to reduce the complexity of a circuit by an examination of the function implemented in the code. A simple example is probably the best way to appreciate this. Take the VHDL code fragment

```
IF a = I THEN
    c <= a * b;
ELSE
    c <= a * d;
```

In this code there are two multiply operations. Yet it is clear by inspection that only one of the multiply operations is active at any one instance. Clearly it would be wasteful to implement two large multipliers when only one is needed. This is what is meant by resource sharing. By creating a single multiplier with a permanent input 'a' surrounded by selector logic to select input 'b' if a = 1 or 'd' otherwise would result in a smaller circuit. The reader will have noticed that alternative resource sharing is possible in this example as if a =1 then c = b.

Resource sharing will have an impact on speed (owing to the extra multiplexing logic), but it is usually worthwhile to include it as an optimization process. The disadvantages of resource sharing include possible slower operating speed and an increase in the time taken to perform the optimization task.

2.4.4 VHDL coding style

The way in which the VHDL is coded can have substantial impact on the way in which the circuit is transformed into a net-list by the synthesizer. The following chapters address this matter in some detail. For example, the CASE statement has an OTHERS option, which says 'if none of the previous case conditions are true, then undertake the action stated in the OTHERS clause'. Including such a clause in a CASE statement can result in substantial area simplification. Further optimizations are possible by judicious use of brackets to indicate an efficient ordering of evaluation. The dos and don'ts sections of the following chapters explore this issue in some depth.

2.5 TECHNOLOGY MAPPING

Sooner or later a design needs to be mapped into a technology net-list and handed over to the place and route software that will provide the final design. Clearly, there is an interaction between these two issues. It is perfectly possible to delay the choice of technology until the synthesis is complete. This occurs throughout the synthesizer, producing a generic technology output, which comprises a set of gates and flip flops with particular delays. This generic description can then be transformed into a particular technology by the use of a technology mapper. This approach has the advantages of delaying technology until a later stage, enabling more time for negotiation with device fabricators. Furthermore, it is possible quickly to change designs into another or even many other technologies, eliminating dependence on a supplier.

Inevitably, there is a price to be paid, and that is one of efficiency. Knowing the technology chosen at a particular stage can aid the optimization software to produce better results. Furthermore, technology mappers are not always perfect pieces of software and they may fail to map from generic to a particular technology. When this problem is faced there is no other choice but to resynthesize the offending circuit directly into the technology, which may be a complex task depending on the size of the circuit.

3

Principles of VHDL

3.1 THE VHDL DESIGN HIERARCHY

Any hardware design can be described in terms of its operation at different levels of abstraction, from system through to logic gate. At each level of this hierarchy the overall inputs and outputs remain the same but the functionality of distinct sections becomes clearer. When the local inputs and outputs and the function of a block are sufficiently defined, the hardware can be designed. The method chosen to achieve this implementation will depend on many factors and this topic is a whole book in itself!

VHDL is capable of describing a well-defined hardware block at any level of abstraction. A **design entity** is the VHDL representation of such a block and can be considered to be at the top of the design hierarchy. Within the design entity, the function of the hardware is often further decomposed by using **external** and **internal blocks**.

External blocks are other design entities that have been previously compiled and stored in a library (section 3.4). These types of blocks are generally referred to as **components**. Each component can further extend the hierarchy by also using external and internal blocks to describe its own function. The present design entity can also be thought of as an external block.

Internal blocks are self-contained functional units that have explicitly defined input and output signals. They therefore represent the next level down in the hierarchy. Again, an internal block may contain external or other internal blocks. The Block statement is used to create an internal block.

A design entity contains an entity declaration and an architecture body. These are described in detail below. Figure 3.1 shows the organization of the VHDL design hierarchy and illustrates the main elements of an architecture body. The **port** connections, indicated by arrowheads, enable information to be passed between blocks and, at the top level, allow the design entity to communicate with the external environment. A design entity may contain any or all of the elements shown in the diagram. It will always have an entity declaration and architecture body at the top level.

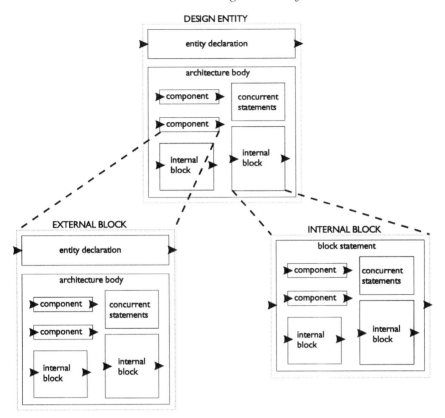

Figure 3.1 Hierarchical organization of a VHDL design.

3.1.1 The entity declaration

The entity declaration is the interface between the external environment, such as a top-level schematic, and the design. It is at the top of the design tree for every external block. Hence, it will usually contain a description of the inputs to and outputs from the block in the form of a port statement. An entity declaration without a port statement does not have any external connections.

The Generic statement can be used within the entity declaration to pass timing, control or environmental data into the design unit. A component declaration may also have a Generic statement. This information is useful when simulating a design, but when synthesizing most of it is irrelevant. Hence, the Generic statement usually only supports the declaration of integer **generics**, which have a constant value and can therefore be substituted into the associated design unit at compile time. The syntax of the entity declaration is described in Box 3.1.

Box 3.1 How to construct the entity declaration.

The entity declaration statement defines the interface to a design unit and the external environment. Apply the following syntax:

```
entity_declaration ::=
    entity entity_identifier is
        entity_header
        entity_declarative_part
    end [ entity ] [entity_identifier ];
```

1 The entity identifier is a simple name. It must be unique among the entities contained in the same design library.
2 The entity header can contain a generic statement. For synthesis, this defines constants that can be accessed by the port statement or the architecture body. A generic list (Chapter 5) is contained inside the generic statement, which has the syntax:

```
generic( identifier{,identifier}:subtype_indication := static_expression{ ;
         identifier{,identifier}:subtype_indication := static_expression });
```

Each identifier is the name of a generic constant in the design. The value of each generic, determined from the static expression, must be included for synthesis purposes (Box 3.3).
3 The entity header can contain a port statement. This defines the I/O structure of the entity using a port list (Chapter 5). This has the syntax:

```
port (identifier{, identifier}:mode subtype_indication [ := expression ]{ ;
      identifier{, identifier}:mode subtype_indication [ := expression ]} );

    mode ::= in | out | inout | buffer
    subtype_indication ::= type_name | subtype_name [ constraint ]
```

Each identifier is the name of a port in the design. It is treated as a signal. The constraint, if any, applied to the type or subtype must be static. It may contain generics defined in a preceding generic statement. The default initial value expression of the port is ignored by the synthesizer.
4 The entity declarative part can include declarations for any of the following items:

subprograms, components, types, subtypes, constants, signals, shared variables, attributes.

It may also contain use clauses, subprogram bodies and attribute specifications. A number of other items are not supported for synthesis. All items in the declarative part are also visible to any architecture that uses the entity. Each declaration is discussed in this or later chapters.

5 An optional entity statement part (preceded by the keyword **begin**) after the declarative part can be included but will be ignored by the synthesizer.

3.1.2 The architecture body

The architecture body is used to specify the relationship between the inputs and outputs declared in the entity. It therefore describes the actual function of the hardware. Although each entity must be unique, several architectures can be associated with one entity. This allows the function of a block to be changed without changing its external structure. The syntax of the architecture body is detailed in Box 3.2.

The architecture body contains any number of concurrent statements – components and internal blocks are also of this type. These statements can be considered to be executing asynchronously and completely independently of each other. It is therefore crucial to understand how the relationships between different concurrent statements are going to affect not only the behaviour but also the logical structure of the hardware that synthesis will produce.

Additionally, certain concurrent statements may contain a number of sequentially executing statements. Such flexibility allows the architecture

Box 3.2 How to construct the architecture body.

The architecture body contains the description of the function of a circuit block. It has the following syntax:

```
architecture body ::=
    architecture architecture_identifier of entity_identifier is
        architecture_declarative_part
    begin
        architecture_statement_part
    end [ architecture ] [architecture_identifier ];
```

1 The architecture identifier is a simple name. It must be unique among the architectures that share the same entity.
2 The architecture declarative part serves the same function as the entity declarative part. However, all declarations are local to the architecture and are not visible to any other architecture blocks.
3 The architecture statement part contains a series of concurrent statements that describe the function of the block in a structural, dataflow or behavioural style, or some combination of these.

body to be constructed using three possible language styles. Choosing a particular approach does not limit the design to that one style. In fact, many functional descriptions can be implemented in more than one style. In general, a combination of two or all three styles is often required to obtain the most compact and clear VHDL code.

Dataflow style

A dataflow style architecture models the hardware in terms of the movement of data over continuous time between combinational logic components such as adders, decoders and primitive logic gates. It describes the register-transfer level behaviour of a circuit. The language topics that are most relevant to the dataflow style of architectures include the following:

- operators – logical, relational and mathematical;
- operator overloading;
- concurrent assignment statements.

This style is not appropriate for the modelling of sequential logic. Instead, and as its name suggests, it is best applied in the modelling of data driven circuit elements such as an arithmetic logic unit (ALU).

Behavioural style

The behavioural style architecture contains concurrent statements with sections of sequential statements that describe the outputs of the circuit at a discrete moment in time given particular inputs. While similar language constructs are often found in dataflow and behavioural style architectures, only the latter explicitly exhibit the notions of time and control. This style describes the function of the circuit at the algorithmic level.

The aspects of VHDL that are most relevant to behavioural style architectures include the following:

- process statements and sensitivity lists;
- sequential statements;
- variables.

This architectural style is used to describe both sequential and combinational circuits. Hence, it is a valuable design approach for finite state machines (FSMs) or any control logic.

Structural style

A structural style architecture describes the circuit primarily in terms of components. These may be drawn from libraries supplied by ASIC vendors, modules that are purposely produced for this design or general, user-defined modules that are stored in a library created for a previous design. The choice of whether to use vendor-specific or generic / user-defined components is an issue. An appropriate balance must be struck between the efficiency of the optimization and technology mapping processes and the future flexibility of the design. This was discussed in Chapter 2.

The architecture body specifies which components are contained in a design and how they are interconnected. The main VHDL topics associated with structural style architectures include the following:

* component declarations and instantiations;
* port mapping and signal interface lists;
* libraries and packages;
* signals (for interconnection).

A structural architecture is often used at the higher levels of a design to enable a clear distinction to be made between the various functional units and enable a hierarchical approach to be adopted in the design process.

3.2 OBJECTS

As with any programming language, VHDL requires a method for storing data or transferring it to another section of the code. An **object** is a named entity or element that contains a value of a given type. There are four classes of object in VHDL – **constants**, **signals**, **variables** and **files**. The final class will be ignored by a synthesizer as it is used to access or store information in the host system environment. The others are discussed below.

An object can be declared in several ways, including any of the following:

* in an object declaration statement;
* in a for statement as a loop or generate statement parameter;
* as a subprogram parameter;
* as a port in or out of a block.

All four methods will create an object that can be referenced by name, but only the first method is an explicit declaration. The other three will create the object implicitly for use only within a restricted section of code, such as a subprogram or loop. The rules governing where a declared object can and cannot be used are similar to those in other programming languages but will be clarified where necessary. The naming of VHDL objects is discussed in Chapter 4.

Note that an object declared in a For statement can be treated as a constant. The objects declared in a port statement are always signals while subprogram parameters could be from any class.

Objects can be used in many types of expression, but care must be taken with some to ensure that the code can be synthesized (Box 3.3).

Box 3.3 How to use static and dynamic objects or values in expressions.

When a description is being compiled, the synthesizer must know that certain values within the design are not going to change, no matter what happens to the inputs. These static values are ones that will directly affect the hardware that the synthesizer will produce. An error will occur if the compiler cannot determine how to construct the hardware because a parameter it requires is dynamic and may change over time.

A prime example is the range of a For loop. Unless the compiler is given static range parameters, it cannot determine how many times to unroll the loop and therefore how many copies of the hardware to construct. A simulator is much less concerned with such issues as it will take the present values of the range parameters and repeat the loop as many times as they specify.

A constant is a static object and so may be used in such a case. Also, integer and based literals (Chapter 6) and enumeration literals are static values. Signals and variables are dynamic objects.

In the following chapters, the requirement for static values in expressions has been indicated where necessary.

3.2.1 Constants

A constant is an object with a static value of a given type. Once the value has been assigned to a constant it cannot be changed. How a constant is represented in the synthesized hardware will be determined by how it is used. It will, of course, be encoded as a binary value when implemented in hardware, but may be declared as any synthesizeable type. Box 3.4 presents the required syntax.

3.2.2 Variables

A variable is an object of any synthesizeable type that can be used as temporary storage within a VHDL description. Depending on where it is declared, a variable can be shared between different concurrently executing statements in an architecture or can be local to one. Box 3.5 explains how and where to declare a variable.

The value stored in a variable can be freely read or updated, and as with other programming languages, any modification is effective immediately.

Box 3.4 How to declare a constant.

A constant is an object with a static value of a given type, declared with the following syntax:

constant_declaration ::=
    constant identifier {, identifier} : subtype_indication [:= static_expression];
subtype_indication ::= type_name | subtype_name [constraint]

1 Each identifier is a simple name.

2 A constant cannot be a file type, access type or a composite type with a subelement that is either a file or access type. Furthermore, these and other types are not supported in the synthesis environment.

3 The expression specifies the value of the constant. It is only optional when making the constant declaration inside a package declaration. In this case, the value of the **deferred constant** must be assigned in the associated package body.

Examples

    constant step : INTEGER := 3;
    constant terminal : STANDARD.BIT_VECTOR(4 downto 0) := "00100";

This example explicitly names the package (**STANDARD**) from which the array type (**BIT_VECTOR**) has been selected (section 3.4).

A variable that is local to a process will be discarded and its value lost once the execution of the process is completed. When the process is restarted the variable will be declared again. It is useful for simulation to assign an **initial value** to a variable when it is declared. If one is not assigned, a default value will be given. Such information is meaningless when design is synthesized and will therefore be ignored. However, ensuring that variables and signals are correctly initialized is an important issue when designing VHDL for synthesis. Failure to assign a value to a variable before it is read will produce unwanted feedback in a design. Refer to Chapter 5 for further elaboration on this point.

A value is assigned to a variable by using the variable assignment symbol ':='.

Shared variables must be used with caution to ensure that multiple assignments to the same variable in different processes are correctly synchronized. There must be no possibility that two processes could be updating the same variable concurrently. This can lead to a design with unpredictable (non-deterministic) simulation and synthesis results.

Box 3.5 How to declare a variable.

A variable is an object that can be used for temporary and immediate storage. Its declaration has the syntax:

variable_declaration ::=
    [**shared**] **variable** identifier {, identifier} : subtype_indication [:=
                                                            expression];

subtype_indication ::= type_name | subtype_name [constraint]

1 Each identifier is a simple name.
2 A **shared variable** must be declared in the declarative part of the entity or architecture. It can be accessed by any process in the architecture. A **variable** must be declared inside a process and is only visible to that one process.
3 The optional expression gives an initial simulation value to the variable. If this is omitted, a default value will be assigned. The synthesizer will ignore the expression.

3.2.3 Signals

The VHDL abstraction of a voltage level on a wire is a signal. A signal therefore encompasses the notion of time and delays. It is an object of any synthesizeable type that can be shared by a number of concurrent statements. It can be created explicitly or implicitly, using one of the methods discussed in section 3.2.

A signal declaration statement creates the object explicitly (Box 3.6). Commonly, the declaration is made in the architecture declarative part, allowing the signal to be visible to any concurrent statement within that architecture.

Implicit declarations may place restrictions on how a signal can be used. For example, the port statement requires that each signal declaration it contains also specifies the direction or **mode** of the port – **in**, **out**, **inout** or **buffer**. The following limitations apply for each mode.

- A signal of mode **in** can be read but not updated.
- A signal of mode **out** can be updated but not read.
- A signal of mode **inout** or without a mode (explicitly declared) can be both read and updated.
- A signal of mode **buffer** can be read but can only be updated by one source.

Normally a signal will be driven by only one source but could drive several inputs. However, in hardware, a signal on a bus, for example, can

Box 3.6 How to declare a signal.

> A signal is an object which has characteristics similar to a signal on a wire. It is declared by apply the syntax:
>
> signal_declaration ::=
>   **signal** identifier {, identifier} : subtype_indication [:= expression];
>   subtype_indication ::= type_name | subtype_name [constraint]
>
> 1 Each identifier is a simple name.
> 2 A signal cannot be a file type, access type or a composite type with a subelement that is either a file or access type. Furthermore, these and other types are not supported in the synthesis environment.
> 3 The optional expression gives an initial simulation value to the variable. If this is omitted, a default value will be assigned. The synthesizer will ignore the expression.

be connected to several sources, each of which can have different drive strengths, including high impedance or resistive pull-up. In this case, **resolution functions** are used to create special **resolved type** signals. The function examines the logic level and strength of each source to determine the appropriate level and strength for the combined signal. However, the use of resolved signal types is only possible when simulating a description. The synthesizer will simply ignore any resolution functions built in to a signal's type definition and wire together all of its sources.

Updating the value of a signal must be performed with care. A signal assignment statement only *schedules* the update for some unspecified time in the future. This is different to a variable assignment statement, which ensures that the update is effective immediately. It is very important to appreciate this difference, especially when making assignments within processes.

As with a variable, the synthesizer will ignore any initial value assigned to the signal. A value is assigned to a signal by using the signal assignment symbol '<='.

3.3 TYPES AND SUBTYPES

Each object used in a VHDL description must be declared as a certain type or subtype. This will place restrictions on the value an object can contain and how the object can be manipulated. A type can be created that satisfies the requirements of a design, but a range of predefined types do exist. These are all stored in the **STANDARD** package within the STD library. Associated with each of these are a full set of standard operators (logic, relational, etc.).

A restriction placed on the range of values that a given type may take is called a **constraint**. A **subtype** is any type with a constraint. Declaring an object as a type subjected to a constraint is equivalent to declaring a subtype. An *unconstrained* type does not have any restriction placed on the range of values that an object of that type may take. The four classes of types are:

- scalar
- composite
- access
- file.

Access and file types are not supported in the synthesis environment and will not be discussed. Additionally, a new type must always be declared in full, immediately – **incomplete type declarations** are ignored by the synthesizer.

The naming of VHDL types is discussed in Chapter 4.

3.3.1 Scalar types

There are four **scalar** types in VHDL – enumeration, integer, physical and floating point. Only the enumeration and integer types are supported by the synthesis environment. For simulation purposes, **physical** types are used to represent measurements of some quantity such as time or distance, while **floating point** types allow the use of real numbers.

Enumeration types

An **enumeration** type is a set of identifiers or characters that each represent a unique numeric value. The following predefined enumeration types are stored in the STANDARD package.

```
type BOOLEAN is (FALSE, TRUE);
type BIT is ('0','1');
type CHARACTER is (
    NUL,  SOH,  STX,  ETX,  EOT,  ENQ,  ACK,  BEL,
    BS,   HT,   LF,   VT,   FF,   CR,   SO,   SI,
    DLE,  DCI,  DC2,  DC3,  DC4,  NAK,  SYN,  ETB,
    CAN,  EM,   SUB,  ESC,  FSP,  GSP,  RSP,  USP,
    ' ',  '!',  '"',  '#',  '$',  '%',  '&',  ''',
    '(',  ')',  '*',  '+',  ',',  '-',  '.',  '/',
    '0',  '1',  '2',  '3',  '4',  '5',  '6',  '7',
    '8',  '9',  ':',  ';',  '<',  '=',  '>',  '?',
    '@',  'A',  'B',  'C',  'D',  'E',  'F',  'G',
    'H',  'I',  'J',  'K',  'L',  'M',  'N',  'O',
    'P',  'Q',  'R',  'S',  'T',  'U',  'V',  'W',
```

'X', 'Y', 'Z', '[', '\', ']', '^', '_',
' ', 'a', 'b', 'c', 'd', 'e', 'f', 'g',
'h', 'i', 'j', 'k', 'l', 'm', 'n', 'o',
'p', 'q', 'r', 's', 't', 'u', 'v', 'w',
'x', 'y', 'z', '{', '|', '}', '~', DEL);

Others also exist but are not defined for use in synthesizeable descriptions. Note that when numerals are used as identifiers they must be written as '0' and not simply 0, as this is an integer value and not an enumerate literal.

Each enumeration literal represents a different value, and this value is determined by the position of the literal in the set. The first in the list is assigned the value 0, the second 1, and so on. In the synthesizer, these values are of course implemented in their binary form. Hence, an enumeration type containing four literals is represented by two bits, but five literals would require three bits.

It is also possible to define your own encoding scheme through the use of the ENUM_TYPE_ENCODING attribute. See Chapter 7 for more details on attributes. Box 3.7 illustrates the syntax of this type of declaration.

Integer types

Integers are the only numeric type supported for synthesis. An **integer** type is a set of integer values between a specified range. The upper and lower bounds of the range are vendor specific, but for fully portable code the limits specified in the predefined integer type INTEGER should be used, i.e.

type INTEGER **is** –2147483647 to 2147483647;

Box 3.7 How to declare enumeration types.

Apply the following syntax:

type_declaration ::=
    type identifier is (enumeration_literal {, enumeration_literal});

1 The enumeration literal can be an identifier (e.g. RED, Blue, green) or a character literal (e.g. '1', 'X', '?'). Each must be unique within the type declaration. Identifiers (simple names) are discussed in detail in Chapter 4.
2 The same enumeration literal can appear in more than one type declaration. In such cases it may be necessary to clarify the type being used in a particular context. A **qualified expression** can be used for this purpose.

An integer type object can be used with any of the standard operators, but certain rules must be observed in order to create a synthesizeable description. These are discussed in section 6.1.2.

The synthesizer will use an appropriate binary representation for each integer type object. Hence, an unconstrained object of type INTEGER requires a 32-bit vector to support its range. It is therefore good design practice always to use integer types with range constraints. New types can be created (Box 3.8) or a constrained INTEGER object type declared. For example:

> **type** SHORT **is range** −128 **to** 127;
> **type** MAGNITUDE **is** 0 **to** 127;
> **signal** A : INTEGER **range** −128 **to** 127; -- 8-bit (1+7) signed number
> **signal** B : SHORT;   -- 8-bit (1+7) signed number
> **signal** C : MAGNITUDE;   -- 7-bit unsigned number

The two integer signals, A and B, are identical. Notice that the type SHORT is **signed** and MAGNITUDE is **unsigned**. This occurs automatically due to the ranges specified in their definitions. When synthesized, an unsigned integer will require simpler resolving logic than a signed one and should therefore be used if possible. Further details of signed and unsigned types can be found in Chapter 6.

Box 3.8 How to declare integer types.

> Apply the following syntax:
>
> type_declaration ::=
>   **type** identifier **is range** simple_expression **to**|**downto** simple_expression;
>
> 1 Each simple expression must return an integer value that is between the bounds of −2147483647 and +2147483647. The attributes 'low and 'high are commonly used to help define the range (Chapter 7).
> 2 If the first simple expression contains the smaller integer value of the two – an ascending range direction – then the keyword **to** is used. If a descending range direction is specified, the keyword **downto** is employed.
> 3 Different ways of expressing integers in VHDL – integer and based literals – are discussed in Chapter 6.

3.3.2 Composite types

A composite type represents a collection of objects that are defined as either composite types or scalar types. These objects can be organized in two ways: as an array or as a record. The objects in a composite type cannot be defined as either access or file types if the type is to be synthesized.

Array types

An **array** type is composed of a number of scalar or composite objects all of the same type. Declaring an array is a convenient way of grouping objects together or, for a signal array, defining a bus. The elements of an array can be accessed individually, as a subset or as a complete group. Most synthesizers will support arrays up to two dimensions. The syntax for their declaration is shown in Box 3.9.

An array can be declared to be either constrained or unconstrained. A **constrained array type** has its range indices specified when the type is declared. All objects of this type then declared will have the same length. An **unconstrained array type** does not have the range indices fixed by the type declaration stage. Instead, they are only specified when an object of that type is declared. This means that objects of an unconstrained array type may all be different lengths. A two-dimensional array may have one or both pairs of indices initially unconstrained. Range indices must be static.

The STANDARD package contains the definition for the array types STRING and BIT_VECTOR. Only the latter is supported when synthesizing and is defined as:

type BIT_VECTOR **is array** (NATURAL **range** <>) **of** BIT;

BIT_VECTOR is an example of a one-dimensional, unconstrained array type. In this and all array type declarations, three parameters must be specified – the type name, the range indices and the type of the array elements. In an unconstrained type, the only part of the range indices that is specified is their type. For example, the indices supplied when defining an object of type BIT_VECTOR must be integers, greater than or equal to 0 – the definition of the type NATURAL (section 3.3.3). The **box** symbol '<>' indicates that the range is presently unspecified.

It is an error to declare an object of an unconstrained type without specifying its dimensions. The exception to this rule is the declaration of a constant. In this case, the value of the constant implicitly defines the dimensions of the array. For example:

constant Y : BIT_VECTOR := "101"; -- an implicitly defined range of 0 to 2

As an array type can be declared to be of a composite type. It is possible to create an array of an array. Hence, there are two alternative ways of

Box 3.9 How to declare array types.

An array can be declared as constrained or unconstrained by applying the syntax:

```
type_declaration ::=
    type identifier is
constrained_array_definition|unconstrained_array_definition;

constrained_array_definition ::=
    array (discrete_range [ , discrete_range ] ) of subtype_indication

discrete_range ::=
    [ type_mark range ] simple_expression to|downto simple_expression

type_mark ::= type_name|subtype_name
subtype_indication ::= type_mark [ constraint ]

unconstrained_array_definition ::=
    array (type_mark range<> [ ,type_mark range<> ] ) of
subtype_indication
```

1 Each simple expression must be static. The attributes 'low and 'high are commonly used to help define the range (Chapter 7).
2 If the first simple expression contains the smaller value of the two – an ascending range direction – then the keyword **to** is used. If a descending range direction is specified the keyword **downto** is employed.
3 The range values of an unconstrained array are defined when an object of that type are declared. The definition of this range is constrained by the type mark. For example, for the subtypes:

subtype SMALL **is** INTEGER **range** 0 to 255;
type MEMORY **is array** (SMALL **range** <>) **of** BIT;

only the first object declaration below is legal.

signal A : MEMORY(0 to 31); legal
signal B : MEMORY(32 to 511); illegal, out of upper bound

4 Manipulating arrays, assigning values to arrays and forms of expression for array (bit-string) literals are discussed in Chapter 5.

creating a two-dimensional array type. Examine the following type declarations.

type MEMORY_A **is array**(0 to 16, 7 **downto** 0) **of** BIT;
type ROW **is array** (7 **downto** 0) **of** BIT;
type MEMORY_B **is array** (0 to 16) **of** ROW;

MEMORY_A is a two-dimensional array of a scalar type, while MEMORY_B is a one-dimensional array of a one-dimensional array type. These two arrays are identical in size. However, the individual elements of each must be address in different ways. Chapter 5 discusses how to access and manipulate array objects.

An array type can be defined with ascending or descending range indices. When declaring an array type to represent a number, the common convention is to use a descending range. This means that the most significant bit in the number's binary representation will have the largest index value. For example:

signal Z : BIT_VECTOR(3 **downto** 0);

The most significant bit of this signal is Z(3).

Bit-string literals are a common form of expression for BIT_VECTOR assignments. Ways of expressing these literals are discussed in Chapter 5.

Record types

A **record** type is composed of a number of scalar or composite objects of any type. There are no predefined record types available in the STANDARD package.

Each object declared in a record is known as an element of that record. Like an array, it is possible to create a record that contains elements declared as other records. It cannot be a **recursive** declaration, containing an element of the record type that is actually being declared. Nor can an element be of an unconstrained type. The syntax is given in Box 3.10.

Chapter 7 discuss how to assign values to and access records.

3.3.3 Subtype declarations

A **subtype** is a type with a constraint on the range of values it may take. There are a number of advantages of using a subtype rather than defining a new type. In particular, a subtype can make use of all the functions and operators defined for the type from which it is derived, its **base type**. When synthesizing a design, subtypes are a useful way of reducing the number of bits required to implement an object. An INTEGER type signal, for example, would require 32 bits for an unconstrained implementation. In the STANDARD package two integer subtypes are defined. These are:

subtype NATURAL **is** INTEGER **range** 0 **to** INTEGER'**high**;
subtype POSITIVE **is** INTEGER **range** I **to** INTEGER'**high**;

These declarations make use of the predefined attribute **high**. This is discussed in Chapter 7. The syntax for creating subtypes is given in Box 3.11.

Box 3.10 How to declare record types.

A record can consist of any synthesizeable scalar or composite type.
It is declared using the syntax:

    type_declaration ::=
        type identifier **is record**
           element_declaration;
        { element declaration; }
           end record [identifier];
    element_declaration ::= identifier {, identifier} : subtype_indication

1 Each element declaration represents one or more elements of the
   record. An element declaration with N identifiers is equivalent to
   N independent element declarations.
2 All identifiers within the element declarations must be unique.
3 A record can contain elements declared as integers, enumerates,
   arrays or even other records. The syntax for each of these declaration has already been shown.

Once a subtype has been declared it can itself be used as the base type
for further subtype declarations. Any two subtypes that are ultimately derived from the same unconstrained type can be mixed in one expression.
However, it is important to ensure that the range constraints on both sides
of an assignment statement are matched. For example, if two integer
subtypes and two signals are defined as follows:

    subtype SHORTER **is** INTEGER **range** I **to** I0;
    subtype LONGER **is** INTEGER **range** I **to** 20;
    signal X : SHORTER;
    signal Y : LONGER;

Legal and illegal assignments would be:

    Y <= X; -- legal since X will be within the range of Y
    X <= Y; -- *illegal since Y may be outside the range of X*

The length of array subtypes and types must be matched in the same
way. Any subtype mismatches will be identified by a simulator only when
a run-time error occurs. Some synthesizer will not compile a VHDL description if any mismatches exist, but others might. In either case,
presynthesis simulation should identify such errors.

3.4 DESIGN ORGANIZATION

Design entities are very rarely fully independent functional units. In most
cases they will reference other (external) blocks that represent components
or contain common subprograms and declarations. These might be stored

Box 3.11 How to declare subtypes.

A subtype is a type with a constrained range of values . It is declared with the syntax:

subtype_declaration ::=
    subtype identifier **is** subtype_indication;
    subtype_indication ::= type_name | subtype_name [constraint]

1 If a constraint is not applied the subtype is exactly the same as the type or subtype from which it is being declared.
2 The form of the constraint expression is dependent upon the type or subtype from which the subtype is being declared. The following examples illustrate **range constraints**:

subtype SHORT **is** INTEGER **range** -128 **to** 127;
subtype ALPHANUMERIC **is** CHARACTER ('0' **to** 'Z');
subtype UPPERCASE **is** ALPHANUMERIC ('A' **to** 'Z');

An array of any type may instead have an **index constraint**. For example:

type ROM **is array** (0 to 127, 7 downto 0) of BIT;
subtype CACHE **is** ROM (0 to 31, 7 downto 0);

3 The range direction in the subtype must be the same as the subtype or type from which the new subtype is being declared.
4 If the type or subtype from which a new subtype is declared is a signed integer, the new subtype will also be a signed integer. This applies even if the new subtype's range only includes non-negative integers.

in the same library as the present design or elsewhere, in global libraries. Each library will usually be subdivided into smaller units that contain the subprograms or declarations that relate to particular object types.

Figure 3.2 illustrates the structure that may be typical of any library. A **design file** contains the VHDL textual description (source code) for a functional element of a design. A number of design files may be used to implement separate elements of the same design. Each file may contain one or more units that describe the structure or function of a circuit element. These **design units** may by either:

- **primary units**, which contain an entity declaration or package declaration (configuration declarations are not supported when synthesizing); or
- **secondary units**, which contain an architecture body or a package declaration.

The primary unit must always be compiled and stored in the library before its associated secondary unit. A secondary unit cannot be compiled unless it has an associated primary unit. Design units are compiled into **library units** and stored in the working design library. A library unit may contain references to elements declared in other libraries. If these are not implicitly accessible, external library references must be attached to the unit.

Attaching these external references does not mean that everything in the library will be synthesized into hardware. Only if a component or subprogram is actually used will it be synthesized. Also, some library functions are built in, which means they are not synthesized at all (Chapter 6).

3.4.1 Packages

A **package** is used to store elements of a design that are frequently used or common to several design units. Package declarations and package bodies are themselves design units which can make use of the resources available in other packages.

Common items that are found in packages include type and subtype declarations, functions and components. Types and subtypes that are implicitly accessible to any design are declared in the **STANDARD** package. This package cannot be modified by the user and is stored in the **STD** library along with another, **TEXTIO**. However, the type declarations and procedures that this contains cannot be used if a design is to be synthesized.

Synthesis tool and CAE vendors will generally supply many predefined packages that supplement or enhance the basic types found in the **STAND-ARD** package. For example, many vendors provide packages containing arithmetic, comparison and shift functions associated with the bit array

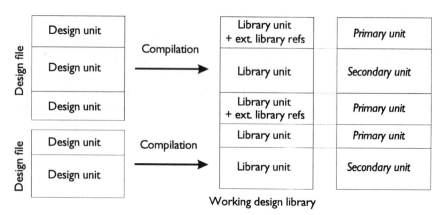

Figure 3.2 Example library structure.

type, BIT_VECTOR. Basic subtypes of this type may also be declared, giving the user access to 4-, 8-, 16- and 32-bit vectors. Other data types, such as std_ulogic and MVL (multivalued logic), are frequently available in vendors' packages. Their use will be discussed in more detail later.

Packages such as these allow the architecture of a circuit to be described in a generic behavioural manner. If a particular technology is being targeted from the start of the design, it is possible to construct a structural description that draws components directly from the packages of the chosen ASIC or FPGA vendor's libraries. This is the best way to ensure that the synthesized design will provide the most efficient implementation of the required function within the chosen hardware.

A package is created with a package declaration. The syntax of this is explained in Box 3.12. Depending on the type of items contained in a package declaration, a package body may not be required. For example, if the package is used only to declare types and subtypes, there is no need for a package body. However, if a component is declared in a package then the description of its behaviour is placed in the body. The syntax of the package body is given in Box 3.13.

3.4.2 Design libraries

Packages, entities and architectures are all stored in a design library. The two types of library are as follows.

- **Resource libraries** generally fall into two types. The first are those that are predefined and provided by synthesis tool or CAE vendors. These primarily contain package declarations and bodies of the sort described in section 3.4.1. The second type are user defined. These may be previous circuit designs which have design units that are common to the present one or simply contain declarations that the designer frequently uses. User-defined libraries may therefore contain not only packages but also entities and architectures.
- There is one only type of **working library**, called WORK. This contains all the design units of the present design that have been compiled. It could therefore store entities, architectures and packages.

The STD library is a resource library that is predefined in the VHDL design environment (not by a particular vendor). Figure 3.3 illustrates the organization of this environment, which consists of two parts – the host environment and the synthesis environment. When a particular design unit or a declaration within one is required by the design unit being compiled, it must be retrieved from the design library stored somewhere within the host environment. The synthesizer must also know where the WORK library is located within the host's environment in order to store the compiled design. Some mechanism for linking logical and physical library

Box 3.12 How to construct a package declaration.

A package declaration is used to store frequently used declarations and subprograms. It has the syntax:

```
package_declaration ::=
    package identifier is
        package_declarative_part
    end [package] [identifier];
```

1 As with an entity's declarative part, the package declarative part may contain declarations for any of the following items:

subprograms, components, types, subtypes, constants, signals, *shared* variables, attributes

It may also contain use clauses (section 3.4.3) and attribute specifications. A number of other items are not supported for synthesis.

2 All items in the package declarative part are visible to any package body that uses the package declaration.

3 Items declared in a package are immediately visible to all subsequent declarations in the same package.

4 The items in a package declaration and package body (if any) are visible to any other design unit which has been given explicit access through a use clause.

Box 3.13 How to construct a package body.

A package body has the syntax:

```
package_body ::=
    package body identifier is
        package_body_declarative_part
    end [package body] [identifier];
```

1 A package only requires a package body if it contains subprograms or deferred constants. These items are located in the package body declarative part.

2 In addition, the following items may be declared in the package body:

types, subtypes, constants, signals, shared variables, attributes

It may also contain use clauses, subprogram bodies and attribute specifications. However, none of these items can be made visible outside of the package body.

3 A package body cannot exist without a package declaration.

4 The identifier is the same as the identifier in the package declaration.

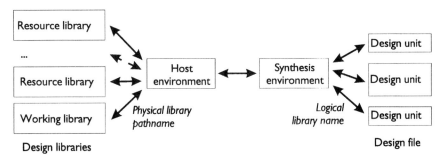

Figure 3.3 The physical and logical environment of VHDL.

names must therefore be provided. VHDL does not define how this is accomplished; this is left to individual CAE tool vendors.

3.4.3 Accessing design libraries and packages

So how can a design unit access a library and its contents? First, there are a number of design units that are implicitly visible to every other design unit. These are the STANDARD package in the STD library and all entities, architectures and packages contained in the WORK library. If these were not implicitly visible, every design unit would need to be preceded by the following statements:

```
library STD,WORK;   -- a library clause
use STD.STANDARD.all;  -- a use clause
```

Second, any library can be made visible to a design unit with a **library clause** as the example above illustrates. Additionally, a **use clause** specifies which packages and particular declarations within a package are required by a design unit. Libraries and packages that are visible to an entity declaration are then implicitly visible to any associated architecture body. The same applies to a package declaration and package body. Hence, the **scope** of a library and a use clause extends from a primary design unit into the associated secondary design unit.

If a use clause is located inside a declarative part of, say, a process, then the specified package contents are visible only to that process. The use clause can be placed in any declarative part to provide this local visibility.

Rather than making the whole package or some of its contents visible to all declarations and/or statements within the scope of the use and library clauses, it is possible to individually and directly access one item in a package. For example, a procedure called DECODE in a package MY_PROCS, in a library MY_LIB could be called using a **selected name** as illustrated in the statement:

```
MY_LIB.MY_PROCS.DECODE(parameter_list);
```

Box 3.14 How to access design libraries and their packages.

To access a particular library, precede the design unit with a library clause. This has the following syntax:

```
library_clause ::=
    library library_logical_name {, library_logical_name};
```

This will *include* the library in the design. To make some or all items contained within a package visible to a design unit, the library containing the package must first be included and then a use clause is required. This has the following syntax:

```
use_clause ::=
    use prefix.suffix;
```

```
prefix ::= library_logical_name.package_identifier
         | library_logical_name
```

```
suffix ::=    item_identifier | all
```

Item identifier is the name of a declaration made within a package. It could therefore be a character literal, simple name or operator symbol.

Examples

To make all declarations in the specified library visible:

```
use library_logical_name.all;
```

To make all declarations in a particular package of the certain library visible:

```
use library_logical_name.package_identifier.all
```

To make a single declaration in a particular package visible:

```
use library_logical_name.package_identifier.item_identifier
```

All package and entity declarations in the working library, **WORK**, and the **STANDARD** package are implicitly visible when the current design unit is being compiled. Any package or library that is visible to a primary unit is implicitly visible to the associated secondary unit.

Items (declaration, subprogram, etc.) can be accessed directly from a library or package without using a library clause or use clause. To access a particular item within a particular package / library, just refer to the item using an appropriate logical prefix (defined above), creating a selected name; viz.:

```
[prefix.]item_identifier
```

A prefix is not required for any items located in the library **WORK**.

In this case, no library or use clauses would be required. The syntax of these clauses is illustrated in Box 3.14.

3.5 SUMMARY

This chapter has provided an overview of the principal elements of the VHDL environment. These can be summarized as follows.

- A hierarchical structure can be created to a design entity through the use of external and internal blocks.
- The entity declaration and architecture body are a pair of design units that enable the function and structure of a circuit to be described.
- The package declaration and package body are also a pair of design units that enable frequently used declarations to be stored in common groups.
- Design libraries enable both vendors and users to store packages, entities and architectures in a structured manner.
- The function of a circuit can be described in one of three architectural styles – structural, dataflow or behavioural.
- Objects are named entities that contain a value of a given data type. The objects constant, variable and signal are available and exhibit very different characteristics.
- An object must be declared to be certain type. Integer and enumeration types are available. These can be formed from a single element – a scalar type – or multiple elements – a composite type. A composite type can be organized as an array or a record.
- Subtypes are types or other subtypes that have a constrained range of values.

It is very important to remember that the VHDL described in this and later chapters is a subset of the language. If a construct cannot be synthesized it is not described in detail but it is always mentioned.

4

Combinational logic design

4.1 BASIC PRINCIPLES OF COMBINATIONAL DESIGN

4.1.1 Which architectural style?

Pure combinational logic primarily implements data-driven elements of a circuit and any simple associated controlling logic. These elements can be anything from 2-bit multiplexers to 16×16 bit multipliers. The architectural style adopted will be determined by, among others, the following factors.

- *The complexity of the function.* Aspects such as its size (bit width) and the regularity of its structure may encourage the use of a particular approach that reduces the work required to actually write code for the design.
- *The level of design abstraction.* If the function can be described in terms of an algorithm, a set of register transfers or even Boolean equations, then the architecture could employ more than one style.
- *The target technology.* If a vendor has been selected, then the designer may wish to use predefined components such as simple logic gates or multiplexers from a vendor's library.
- *The required circuit characteristics.* If the circuit must possess a particular gate level structure, then that must be imposed by the designer using a more structural style. If the structure is flexible and the synthesizer is allowed to select by itself how the circuit is organized, then a more flexible (behavioural) style can be used.

Although it is important to consider all these issues, for most fully data-driven circuit elements the dataflow style is often the most appropriate. These elements are best joined together through a structural higher level in which common components can be easily replicated.

The examples in this chapter will illustrate how all architectural styles can also be applied to a straightforward design.

4.1.2 Logical operators

VHDL defines logical operators for the basic types of BIT and BOOLEAN or any one-dimensional array of either type, such as BIT_VECTOR. The result generated by each operator is of the same type and has the same length as the operands supplied. Hence, the operands must be identical in type and length to allow each matching element to be compared. The synthesizer should detect any errors of this type, but they may go undetected until run time when simulating the description. The operators are listed in Table 4.1.

The Not operator belongs to a class of operators that have the highest precedence. The rest of the logical operators have the equal lowest precedence of all operators. A summary of operator precedence is given in Appendix A.

As the And, Nand, Or, Nor, Xor and Xnor operators are all of equal precedence, parentheses must be used to prioritize the operations when they are mixed or all inverting. As Not has a higher precedence, it does not require parenthesizing. Assuming that all the following signals are of one type, either BIT or BOOLEAN, the following examples illustrate legal and illegal logical expressions.

```
C <= A nand B;  -- legal, parenthesis not required
C <= A and not B;  -- legal, not has a higher precedence
D <= (A nand B) xor C;  -- legal, parenthesis used
D <= (A and not B) xor C;  -- legal, parenthesis used where necessary
D <= A and B and C;  -- legal, one type of non-inverting operator
D <= A nand B nand C;-- illegal, inverting operators need parenthesis
D <= A and B or C;  -- illegal, mixed operators need parenthesis
```

Given the following array objects:

```
signal W,X : BIT_VECTOR(3 downto 0);
signal Y : BIT;
signal Z : BIT_VECTOR(4 downto 0);
```

the following are examples of illegal logical expressions:

```
W <= X or Y;-- mismatch in operand lengths
Z <= W and X;-- operands match each other in length but not the target
```

See section 4.4 for a possible pitfall when using array types with operators.

4.1.3 Relational operators

Relational operators test for equality, inequality and ordering of operands. All the basic synthesizable types in the STANDARD package have available the predefined operations listed in Table 4.2. Operands may be both scalar

Table 4.1 VHDL logical operators

Operator	Function
not	Complement
and	Logical And
nand	Complement of logical And
or	Logical Or
nor	Complement of logical Or
xor	Logical exclusive Or
xnor	Logical exclusive complement of Or

and composite. The ordering operators are not defined for record types. A result of type **BOOLEAN** is always returned. All relational operators have equal precedence, at the level above logical operators. See Appendix A1 for full details of operator precedence.

The operands must be of the same type but may vary in length. Examples of expressions containing scalar operands are:

A <= B>C;-- A is true whenever B is greater than C
F <= D=E **nand** (B=C);-- F is false if D equals E and B equals C

Ordering operations can be performed on operands that are scalars or arrays composed of integer or enumeration type elements. For the 'less than' condition to be **TRUE**, any one of the following must hold.

- The left operand is empty (null) and the right operand is not (non-null).
- The leftmost element of the left operand is less than that of the right.
- The leftmost elements are equal but all remaining elements in the left operand are less than their matching elements in the right operand.

If one operand is shorter than the other then the remaining elements of the longer one are compared with null elements. These are always 'less than' non-null elements. Similar conditions can be applied for the other ordering operators.

Table 4.2 VHDL relational operators

Operator	Function
=	Equality. Returns TRUE if operand values and lengths are exactly equal, otherwise FALSE. Operands can be any synthesizable type
/=	Inequality. Returns TRUE if operand values and/or lengths are different, otherwise FALSE operands can be any synthesizable type
<	'Less than' (see text)
<=	'Less than or equal to' (see text)
>	'Greater than' (see text)
>=	'Greater than or equal to' (see text)

Shift operators are defined for one-dimensional array types. They are discussed in Chapter 5. The predefined arithmetic operators are discussed in Chapter 6.

4.2 4-TO-1 MULTIPLEXER

This example introduces most of the VHDL constructs that are used in combinational logic design. The flexibility of the language is illustrated by showing how the same simple multiplexer can be designed in many different ways using behavioural, structural and dataflow architectural styles, leading to several different synthesized circuits. The following topics are discussed in this section:

- simple names and identifiers
- expressions using integer, based or enumeration literals
- concurrent statements:
    conditional signal assignments
    selected signal assignments
    component instantiations
- sequential statements:
    Case
    If

4.2.1 Functional overview

The 4-to-1 multiplexer selects an output Y from one of four inputs A, B, C or D based on the state of the two select lines S0 and S1. The logical function of the circuit is shown in Table 4.3 and a top-level block diagram showing the inputs and output is illustrated in Figure 4.1.

Hence, the logic function that the architecture must implement is given below.

$$Y = A.\overline{S1}.\overline{S0} + \overline{S1}.S0 + C.S1.\overline{S0} + D.S1.S0 \qquad (4.1)$$

Two-valued logic is assumed, so all signals (and possibly variables) will only occupy the states 0 or 1. Therefore each element must be of type BIT.

Table 4.3 Logic function of the 4-to-1 multiplexer

S1	S0	Output
0	0	A
0	1	B
1	0	C
1	1	D

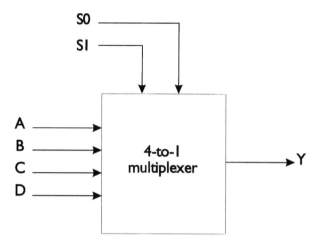

Figure 4.1 Top level of a 4-to-1 multiplexer.

4.2.2 VHDL code design

Library selection

The first stage is to select the libraries that the design is to use. The basic libraries and packages that are available to a VHDL designer have been discussed in Chapter 3. This example uses objects of type **BIT** and **BIT_VECTOR**, and executes basic logic operations. It therefore only requires the **STD** library with its **STANDARD** package. As these are implicitly visible to every design unit, Library and Use clauses are not required.

The entity declaration

The entity describes the top-level structure of a design and acts as the I/O interface between the multiplexer and the external environment. The connections at the top level of the design hierarchy are illustrated in Figure 4.1. The entity declaration required for this circuit is shown in Figure 4.2. It does not contain a Generic statement, default port values or a entity statement part, all of which cannot be synthesized.

The **Port** statement specifies the names, mode and types of the interface signals in the multiplexer. A name given to any item, from a port signal to an architecture body, must follow a certain convention if the design is to synthesize correctly. These limitations are discussed in Box 4.1.

The architecture body

The architecture body describes the function of the multiplexer circuit using a dataflow, behavioural or structural design style. An architecture must

```
entity MUX4TOI is
   port(
          A,B,C,D,S0,SI : in BIT;
          Y : out BIT
          );
end MUX4TOI;
```

Figure 4.2 Entity declaration of the 4-to-1 multiplexer.

be associated with only one entity, although this connection is only made at compile time. Thus, several architecture bodies may reside in one design file along with a single entity. This allows the required architecture body to be selected from several variants depending upon the area, speed and functional requirements. The multiplexer's entity can be associated with one of several architecture bodies, which each illustrate particular language constructs. Each one demonstrates not only the use of the construct but also its circuit characteristics when synthesized.

Dataflow design style
Four architecture bodies have been constructed using concurrent signal assignments in the dataflow style. These are shown in Figure 4.3.

In DATAFLOWI and DATAFLOW2 **unconditional signal assignment** statements are used. As both show, the signal Y is simply assigned the BIT type result of the logical expression. The difference between these two dataflow architecture occurs in their use of parentheses. In both cases it is necessary to use one level of parentheses to ensure that the expression is correctly evaluated. However, the use of multiple levels could force the synthesizer to generate a different combinational circuit. Remember that the Not operator takes precedence over other logical operators, which all have the same level of precedence.

The two architectures illustrate the use of one type of concurrent statement. DATAFLOW3 demonstrates the use of another – the **conditional signal assignment**. This assigns one of N values to a single target, in accordance with the satisfied condition. The condition attached to each branch can be different. Box 4.2 gives a description of its syntax. It is a two-way branch statement that will construct a series of two-input multiplexers. With each branch that is added, another stage in the multiplexer tree is introduced. For example, the statement Z <= A when C=0 else B; will create the circuit shown in Figure 4.4. It is possible to specify explicitly that the signal should not change when a certain condition is true by using the keyword **Unaffected**. In the following example the two branches will assign the same original value to the target, A.

```
A <= A when B=3 else
     unaffected;
```

Box 4.1 How to select and use simple names

Simple names or identifiers provide labels for objects, types, etc. and create associations between discrete units of a design, such as an architecture and an entity.

Basic identifiers are formed from upper- or lower-case letters, digits and the underline character. The first character must always be a letter. The following are examples:

Z   A4   CountUpDown   RST_SET

An identifier of this type cannot be the same as any of the VHDL reserved words that are listed in Appendix A1. Additionally, many synthesizers may also place restrictions on the use of the following:

- common words such as CLOCK, GROUND, INPUT and LOW as object identifiers within a design unit;
- generic style names such as AND2, OR4 and FDxxxxx as identifiers of actual design units (entity name, architecture name, etc.).

The synthesis tool vendor should provide full details of these restrictions in its documentation.

Most synthesizers are case insensitive and so would recognize a signal called Count to be the same as one called COUNT. The synthesizer will convert everything to upper case when compiling.

Caution must be exercised with array names since the synthesizer may strip off their parentheses. For example, an array signal element A(0) would be converted to A0 by the compiler. This will cause an error if a scalar signal has also been named A0. Therefore, avoid using names that end in a number, such as LOAD1 or D6, when also using array types. This also applies to variables and constants.

Two-dimensional arrays and records are also treated in similar way. The synthesizer will probably convert each element in the following way:

THIS_ARRAY(2,3)	becomes	THIS_ARRAY2_3
THAT_ARRAY(2)(3)	becomes	THAT_ARRAY2_3
THIS_RECORD.a	becomes	THIS_RECORD_A_
That_Record.B(4,1)	becomes	THAT_RECORD_B4_1_

This statement does nothing more than assign A to A, but it illustrates the point. The Unaffected keyword is a useful way of binding up any untested conditions that do not affect the value of the target. In this respect, it is similar to a null statement used in a process.

DATAFLOW4 is an example of another concurrent statement, a **selected signal assignment**. This performs a similar function to the conditional sig-

```
-- Entity declaration shown in Figure 4.2
```
-- Concurrent signal assignment
```
architecture DATAFLOW1 of MUX4TO1 is
begin
    -- parentheses required when mixing operators of equal precedence
    Y<=  (A and not S1 and not S0) or (B and not S1 and S0) or
         (C and S1 and not S0) or (D and S1 and S0);
end DATAFLOW1;
```
-- Concurrent signal assignment with additional parentheses
```
architecture DATAFLOW2 of MUX4TO1 is
begin
    -- extra parentheses control the order of evaluation/structure of the
    -- circuit
    Y<=  (A and (not S1 and not S0)) or (B and (not S1 and S0)) or
         (C and (S1 and not S0)) or (D and (S1 and S0));
end DATAFLOW2;
```
-- Conditional signal assignment
```
architecture DATAFLOW3 of MUX4TO1 is
begin
    -- each branch condition may be different
    Y<=  A when (S1='0' and S0='0') else
         B when (S1='0' and S0='1') else
         C when (S1='1' and S0='0') else
         D;
end DATAFLOW3;
```
-- Selected signal assignment
```
architecture DATAFLOW4 of MUX4TO1 is
signal SEL : BIT_VECTOR(1 downto 0);
begin
    -- each branch tests for a different value of the same condition with
    -- SEL select
        Y <= A when "00",
             B when "01",
             C when "10",
             D when "11";
    SEL <= S1&S0; -- '&' concatenation operator, see Chapter 5
end DATAFLOW4;
```

Figure 4.3 Four variants of the 4-to-1 multiplexer in a dataflow design style.

nal assignment statement, but is less flexible and has certain hardware advantages over all the other constructs discussed here. It again assigns one of N values to a single target but, in this case, only one condition is tested for each of the N branches. Again the keyword Unaffected can be used. As

Box 4.2 How to use a conditional signal assignment statement

---

1 This is a concurrent statement that contains a series of two-way branches with the outcome of each one being dependent on a different condition. Apply the following syntax:

conditional_signal_assignment ::=
    target_signal <= expression| **when** condition| **else**
                    {expression2 **when** condition2 **else**}
                    expressionN;

2 The final expression cannot have a condition attached.
3 Each expression must evaluate to the same type as the target signal.
4 The keyword Unaffected can be used instead of any expression to ensure no change to the value of the target signal when a particular condition is true. It is similar to the sequential null statement. See section 4.4 for a possible synthesis pitfall.
5 The output will be assigned a new value on a 'first true, first selected' basis. Conditions evaluation therefore takes the form of a multiplexer tree. See section 4.4 for a pitfall to avoid with this approach.
6 This statement is equivalent to using an If-Else statement inside a process.

---

will be seen shortly, the operation of this concurrent statement can be likened to that of a Case statement inside a process. See Box 4.3 for its syntax.

Note that the ordering of the assignment statements in the **DATAFLOW4** architecture does not affect its operation as the assignments to both Y and SEL are performed concurrently. Any number of signals could be assigned values concurrently in the same architecture. Each one will build up its own block of combinational logic and evaluate its output in a continuous and wholly asynchronous manner. As concurrent assignments occur continuously, the output of one may rely on the output of another. This does not apply to sequential signal assignments inside processes. Chapter 5 discusses the slightly more complex principles of these in more detail. Other types of concurrent statement are introduced below in the behavioural design style architectures.

Behavioural design style
Figure 4.5 illustrates two ways of constructing the multiplexer circuit description using the behavioural design style. The entity associated with both architectures has been redeclared and is also shown. For these examples, a **BIT_VECTOR** type signal S, 2 bits wide, has been declared for the

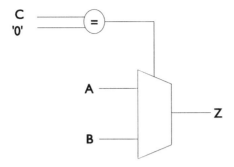

Figure 4.4 Multiplexer created by a conditional signal assignment statement.

Box 4.3 How to use a selected signal assignment statement

---

1 This is a concurrent statement that selects the new target value from the one of N multiplexer outputs that satisfies the single condition under test. Apply the following syntax:

selected_signal_assignment ::=
    with selected_expression **select**
        target_signal <= expression I **when** choice_expression I,
                        {expression2 **when** choice_expression2,}
                        expressionN **when** choice_expressionN|**others**;

2 The choice expression must be of the same type as the selected expression and may take any of the following forms:

value
value I **to** value2   -- a continuous range of values
value I | {value2 | } valueN   -- a set of values

3 A value cannot appear in the list of choice expressions more than once, either explicitly or as part of a range.

4 The **Others** keyword must be used as the final choice of expression if not all possible permutations of the selected expression have been covered by earlier expressions.

5 The target signal and the expressions being assigned must all be of the same type.

6 The keyword Unaffected can be used instead of any expression to ensure no change to the value of the target signal for particular choice expression. This is similar to the sequential null statement. See section 4.4 for a possible synthesis pitfall.

7 This statement is roughly equivalent to a Case statement inside a process.

control/select inputs. Instead of the independent signals S0 and S1, these inputs must be referenced as the array elements S(0) and S(1).

The behavioural style can be identified by the existence of a **Process** statement. A **Process** statement is a concurrent statement that is executed in parallel with conditional or selected signal assignments and all others that will be discussed later. Itself, it will contain a list of statements that execute sequentially, just like any other programming language. A **Process** statement can be thought of as a block that models the behaviour of a particular subsection of a circuit. Typically, it is used when the behaviour is too complex to model with simple signal assignment statement or for describing a synchronous section of a circuit. A deeper discussion of the **Process** statement is reserved for Chapter 5.

The two example architectures in Figure 4.5 illustrate the use of the sequential statements, **Case** and **If-Else**. The **Case** statement demonstrates how useful it is for decoding buses in a simple and clear manner. The construct evaluates an expression (a signal in this case) with a number of alternatives to determine the appropriate sequence of actions. This sequential statement is similar to the selected signal assignment statement in that all possible branches are evaluated in parallel. The **Case** statement is, however, more flexible as each branch may initiate a full sequence of operations, not just a signal assignment. Note that signal assignments that occur inside a process are sequential. These have very different characteristics from concurrent assignments (conditional or selected) used outside the process as mentioned earlier. Chapter 5 deals fully with signals and variables assigned within a process.

All permutations of the expression being evaluated by the **Case** statement must be tested. This example tests all four possible states of S, but in some situations not all of them may have a sequence of actions defined or several may require the same actions. The **Others** keyword can be used in place of the final case branch to group such states together. Often, no action is required and the statement will take the form **when others => null;**. The syntax for the **Case** statement is given in Box 4.4.

The second architecture illustrates an alternative approach that uses an **If-Else** statement. This sequentially evaluates each expression in the construct until it finds one that is true and then executes the associated sequence of actions. This sequential statement is similar to the conditional signal assignment in that each possible two-way branch is evaluated in series, generating a multiplexer tree. Like the **Case** statement, the **If-Else** can initiate any number of actions within each branch, unlike its concurrent equivalent.

Since the expression being evaluated at each stage can be different from the previous one, it is possible to construct an incomplete **If** statement or one that duplicates conditions. As a result, the if-else construct should be applied with caution. As the **IF_ELSE** architecture demonstrates, it is good

```
entity MUX4TO1 is
    port (
            A, B, C, D : in BIT;
            -- Inputs declared using an array type
            S : in BIT_VECTOR(1 downto 0);
            Y : out BIT
        );
end MUX4TO1;
```

-- The Case statement

```
architecture CASE1 of MUX4TO1 is
begin
    process (A,B,C,D,S(1 downto 0))
    begin
        case S is
            -- All states of S explicitly covered by this Case statement.
            -- Array referencing with a bit string, see Chapter 5.
            when "00" => Y <= A;
            when "01" => Y <= B;
            when "10" => Y <= C;
            when "11" => Y <= D;
        end case;
    end process;
end CASE1;
```

--The If-Else statement

```
architecture IF_ELSE of MUX4TO1 is
begin
    process (A,B,C,D,S(1 downto 0))
    begin
        -- Array referencing with an index, see Chapter 5.
        if (S(1)='0' and S(0)='0') then
            Y <= A;
        elsif (S(1)='0' and S(0)='1') then
            Y <= B;
        -- Array referencing with a bit string.
        elsif (S="10") then
            Y <= C;
        -- Implicit coverage of last permutation of S.
        else
            Y <= D;
        end if;
    end process;
end IF_ELSE;
```

Figure 4.5 Two variants of the 4-to-1 multiplexer in a behavioural design style.

Box 4.4 How to use a Case statement

---

1 This is a sequential statement that can appear only inside a process. Apply the following syntax:

```
case_statement ::=
    [label] case case_expression is
        when choice_expression1 => --any sequential statements
        {when choice_expression2 => -- any sequential statements}
        when choice_expressionN|others => -- any sequential statements
    end case [label];
```

2 Each Choice expression must be of the same type as the Case expression and may take any of the following forms:

```
value
value1 to value2            -- a continuous range of values
value1 | {value2 | } valueN  -- a set of values
```

3 A value cannot appear in the list of choices more than once, either explicitly or as part of a range.
4 The Others keyword must be used in place of the final choice expression if not all possible values of the Case expression have been covered by earlier Choice expression.
5 If a label is given at the end of the statement, it must be the same as the label at the start.

---

practice to use a final default branch with no condition applied. This is equivalent to the **'when others =>'** clause in a **Case** statement. The If-Else construct should not be used in preference to a **Case** statement if possible. See section 4.4 of for a further discussion of this and Box 4.5 for the statement's syntax.

CASE1 and IF_ELSE illustrate two ways in which one or more elements of an array can be accessed. S is a simple name that allows the array to be accessed as a bit string, whereas S(1) is an indexed name that allows one particular element to be referenced. Chapter 5 deals with these and other array manipulation techniques.

When a branch condition is satisfied in the architecture CASE1, one of the input signals A, B, C or D is assigned directly to Y. In the architectures of Figure 4.6, two alternative methods of assigning the output are demonstrated. In CASE2, a variable, TEMP, of type BIT has been declared. A private (non-shared) variable, declared within the process declarative part, is always local to the process. The value stored in TEMP after the Case statement has been evaluated must be assigned to Y inside the process. TEMP cannot be passed out of the process or accessed by any other concurrent statement, such as another process. If the variable was shared (Chapter 3)

Box 4.5 How to use an If statement

1 This is a sequential statement that must be used inside a process. Apply the following syntax:

    if_statement ::=
        [label:] **if** condition1 then
            -- any sequential statements
        {**elsif** condition2 then
            -- any sequential statements}
        [**else**
            --any sequential statements]
        end if [label];

2 Elseif branches can be nested to any level but only one **End if** is required.
3 Avoid incomplete **If** statements by including a default **Else** branch. See Chapter 5 for more details of incomplete Ifs.
4 If a label is given to the end of the statement, it must be the same as the label at the start.

then any process could access it. However, the preferred way of passing a value out of a process is to use a signal.

In **CASE3**, TEMP is declared as a signal and so the best place to assign Y is outside the process. In this example, therefore, there are two concurrent statements – the process and the signal assignment. In theory, individual blocks of combinational logic will be created that describe each one. In practice, there is no logic associated with the assignment statement and so the name of the output from the process block is simply changed from TEMP to Y.

Structural design style
Just as a **Process** statement identifies a behavioural style, component declarations and instantiations identify a structural style. Unlike the behavioural style, which specifies what the circuit has to do, a VHDL description in the structural style determines the operation of the circuit by the manner in which any number of functional blocks – components – are interconnected. These may contain anything from simple two-input logic gates to 16×16 bit multipliers.

Three different architecture bodies have been designed in the structural style for the multiplexer example. The components in each one are as follows:

- only N-input Nand gates;
- any type of two-input logic gates or inverters;
- the 'optimum' type of logic gates.

```
-- entity declaration shown in Figure 4.5
-- The Case statement using a variable in the process
architecture CASE2 of MUX4TO1 is
begin
    process (A,B,C,D,S(1 downto 0))
    variable TEMP : BIT;
    begin
        case S is
            when "00" => TEMP := A;
            when "01" => TEMP := B;
            when "10" => TEMP := C;
            when "11" => TEMP := D;
        end case;
        -- The variable is not visible outside of the process.
        Y <= TEMP;
    end process;
end CASE2;
-- The Case statement using a local signal
architecture CASE3 of MUX4TO1 is
    signal TEMP : BIT;
begin
    process (A,B,C,D,S(1 downto 0))
    begin
        case S is
            when "00" => TEMP <= A;
            when "01" => TEMP <= B;
            when "10" => TEMP <= C;
            when "11" => TEMP <= D;
        end case;
    end process;
    -- A concurrent signal assignment is used to pass the result out to Y.
    Y <= TEMP;
end CASE3;
```

Figure 4.6. Using variables and signals with a process.

In each case, the components that are used must be declared and instantiated. A **component declaration** informs the architecture that a particular component is being used and lists its inputs and outputs. The declaration may be located in the architecture's declarative part or a package declaration. It is possible to omit the declaration and instantiate the entity and architecture of the component directly (Box 4.6).

A **component instantiation** statement creates a unique occurrence or instance of a particular component within an architecture. The statement includes an instance name or label, the component's library name (or entity and architecture name) and a list of this particular component's input and output connections. Instantiations of components can only be made inside the architecture and are therefore concurrent statements which, by their nature, have no specific order of evaluation. Note that only signals can be passed in to and out from a component (Box 4.7).

In the first architecture, the only components that have been used are Nand gates with two, three and four inputs. The entity declarations and architecture bodies for each of these very simple components, describing their function in a dataflow style, are shown in Figure 4.7. For this example, these are compiled and stored in the working library, WORK. The architecture which uses these components is STRUCTURAL1, shown in Figure 4.8. This must generate the complements of *S(1)* and *S(0)* using two-input Nand gates. The component instances, INVERT0 and INVERT1, do this by driving both the inputs of each gate with the same signal, either S(0) or S(1), generating the inverted outputs, I0 and I1.

One product term for each of the four inputs must also be determined. For example, the component instance PROD0 is a three-input Nand gate that will output the complement of A if both S(1) and S(0) are low. To generate the final sum term, all the four product terms must be fed into a four-input Nand gate. The Boolean logic for this circuit is shown in equation 4.2.

$$Y = A.\overline{S1}.\overline{S0} + B.\overline{S1}.S0 + C.S1.\overline{S0} + D.S1.S0$$
$$Y = \overline{(A.\overline{S1}.\overline{S0}).(B.\overline{S1}.S0).(C.S1.\overline{S0}).(D.S1.S0)} \tag{4.2}$$

Box 4.6 How to make component declarations

1 A component declaration provides an architecture with the interface characteristics of a component that it may then instantiate. The syntax is as follows:

   component declaration ::=
      component component_identifier [is]
         port (port_list);
      end component [component_identifier];

2 The declaration can be located in the architecture's declarative part (before the keyword **Begin**) or a package which the architecture can access through a **Use** statement.

3 The port list should match the list given in the component's entity declaration. Constructing a port list is discussed in Chapter 5.

4 Any entity–architecture pair can be declared as a component as long as the library is visible to the design unit.

Box 4.7 How to make component instantiations.

1　A component instantiation is a concurrent statement that creates a single occurrence of a particular component. Use the following syntax:

component_instantiation ::=
　　label : instantiate_unit **port map** (signal_list);

instantiate_unit ::=
　　[**component**] component_identifier |
　　entity entity_identifier [(architecture_identifier)]

2　All labels within a particular architecture must be unique. Labels may be repeated on different levels of the design hierarchy.
3　The signal list must contain the same number of signals as the port declaration. If a port on the component is to be left unconnected, the keyword **Open** must be substituted for the signal name.
4　The instantiate unit specifies the method by which the component is being instantiated. If it is being instantiated after it has been declared (Box 4.6) then the component identifier is specified. If no declaration exists then the component is being directly instantiated and the entity_identifier with its compulsory keyword **Entity** are specified.
5　When using the direct method, the library containing the entity and architecture must be made visible to this design unit by using a **Library** clause.
6　The architecture identifier is optional when directly instantiating a component. If it is provided, the name must match that of the secondary unit (the architecture) associated with the specified primary unit (the entity) in the instantiation statement. If it is not provided, the synthesizer will look in the WORK and any other visible libraries in sequence for the specified entity. The architecture associated with that entity will be automatically selected. This may lead to the incorrect component behaviour if the same entity identifier occurs in several different libraries.

As the comments in Figure 4.8 explain, because the components are compiled and stored in the working library, they are automatically visible to this architecture. STRUCTURAL1 demonstrates the use of both component declarations and indirect instantiations.

The second structural style architecture makes use of only two-input And gates, two-input Or gates and inverters. As a result, more component instances and so more component interconnection is required. In this instance, the equation for the logic of the function is equation 4.3.

```
-- 2 input Nand gate
entity NAND_2 is
   port (
           A, B : in BIT;
           C : out BIT
           );
end NAND_2;
architecture DATAFLOW of NAND_2 is
begin
   C<= not(A and B);
end DATAFLOW;
-- 3 input Nand gate
entity NAND_3 is
   port (
           A, B, C : in BIT;
           D : out BIT
           );
end NAND_3;
architecture DATAFLOW of NAND_3 is
begin
   D<= not(A and B and C);
end DATAFLOW;
-- 4 input Nand gate
entity NAND_4 is
   port (
           A, B, C, D : in BIT;
           E : out BIT
           );
end NAND_4;
architecture DATAFLOW of NAND_4 is
begin
   E<= not(A and B and C and D);
end DATAFLOW;
```

Figure 4.7 Nand gate dataflow descriptions. These are compiled and stored in the WORK library.

$$Y = ((A.(\overline{S1}.\overline{S0}) + (\overline{S1}.S0))) + ((C.(S1.\overline{S0})) + (D.(S1.S0))) \tag{4.3}$$

The entity declaration and architecture body of each component is shown in Figure 4.9. Again these descriptions are compiled and stored in the working library, where the multiplexer architecture, STRUCTURAL2, can access them. Unlike STRUCTURAL1, however, this architecture does not declare the components in its declarative part. Instead it uses a package declaration to store the component declarations. A Use clause is then required to

```
-- Entity declaration shown in Figure 4.5
-- No library or use clause indicates that the synthesized components
-- must be located in the WORK library.
-- Nand gate 4 to 1 multiplexer
architecture STRUCTURAL1 of MUX4TO1 is
    -- Component declarations in the architecture declarative part
    component NAND_2
        port (A,B : in BIT; C : out BIT);
    end component;
    component NAND_3
        port (A,B,C : in BIT; D : out BIT);
    end component;
    component NAND_4
        port (A,B,C,D : in BIT; E : out BIT);
    end component;
    signal I0,I1,I2,I3,I4,I5 : BIT;
begin
    -- Generate inverted control signals.
    INVERT0: NAND_2 port map (S(0), S(0), I0);
    INVERT1: NAND_2 port map (S(1), S(1), I1);
    -- Generate each product term.
    PROD0: NAND_3 port map (I1, I0, A, I2);
    PROD1: NAND_3 port map (I1, S(0), B, I3);
    PROD2: NAND_3 port map (S(1), I0, C, I4);
    PROD3: NAND_3 port map (S(1), S(0), D, I5);
    -- Generate sum term.
    SUM: NAND_4 port map (I2, I3, I4, I5, Y);
end STRUCTURAL1;
```

Figure 4.8 4-to-1 multiplexer using only Nand gates.

make the package visible to the multiplexer's entity declaration and architecture body.

This sounds a little complicated, but it is a useful way of tidying up the architecture and of keeping like declarations together. The package, called SIMPLE_COMPS, is shown in Figure 4.10. This contains exactly the same declarations that would have occurred in the architecture's declarative part. This package is compiled and stored in the working library. Figure 4.11 contains the description of the multiplexer. As the package is stored in the implicitly visible library WORK it is only necessary to instruct the compiler to use all the declarations in the package SIMPLE_COMPS with a Use clause. This generally precedes the entity declaration. The architecture statement part is unchanged.

```
-- 2 input AND gate
entity AND_2 is
   port (
         A, B : in BIT;
         C : out BIT
         );
end AND_2;
architecture DATAFLOW of AND_2 is
begin
   C<= A and B;
end DATAFLOW;
-- 2 input OR gate
entity OR_2 is
   port (
         A, B : in BIT;
         C : out BIT
         );
end OR_2;
architecture DATAFLOW of OR_2 is
begin
   C<= A or B;
end DATAFLOW;
-- inverter
entity INVERT is
   port (
         A : in BIT;
         B : out BIT
         );
end INVERT;
architecture DATAFLOW of INVERT is
begin
   B <= not A;
end DATAFLOW;
```

Figure 4.9 Simple gate dataflow descriptions. These are compiled and stored in the WORK library.

The third architecture uses the gates shown in Figure 4.12 that are of the 'optimum' type and fan-in to implement the multiplexer most efficiently. The gates that were chosen for this implementation were three-input And gates, four-input Or gates and inverters. The entity declarations and architecture bodies of the first two are shown in the figure, but the inverter's description was contained in Figure 4.9. All three synthesized components are stored in a new library called GATES.

```
-- No library or use clause indicates that the components
-- declared in the package must also be stored in the WORK library.
package SIMPLE_COMPS is
    component AND_2 is
        port (A,B : in BIT; C : out BIT);
    end component AND_2;
    component OR_2 is
        port (A,B : in BIT; C : out BIT);
    end component OR_2;
    component INV is
        port (A : in BIT; B : out BIT);
    end component INV;
end package SIMPLE_COMPS;
```

Figure 4.10 Package compiled and stored in the WORK library.

```
-- Include the package containing the component declarations
use WORK.SIMPLE_COMPS.all;
-- Entity declaration shown in Figure 4.5
-- Simple logic gates
architecture STRUCTURAL2 of MUX4TO1 is
    signal I0,I1,I2,I3,I4,I5,I6,I7,I8,I9,I10,I11 : BIT;
begin
    -- generate inverted control signals
    INVERT0: INV port map (S(0), I0);
    INVERT1: INV port map (S(1), I1);
    -- generate partial product terms for select signals
    PROD0: AND_2 port map (I1, I0, I2);
    PROD1: AND_2 port map (I1, S(0), I3);
    PROD2: AND_2 port map (S(1), I0, I4);
    PROD3: AND_2 port map (S(1), S(0), I5);
    -- generate full product terms
    PROD4: AND_2 port map (I2, A, I6);
    PROD5: AND_2 port map (I3, B, I7);
    PROD6: AND_2 port map (I4, C, I8);
    PROD7: AND_2 port map (I5, D, I9);
    -- generate sum term
    SUM0: OR_2 port map (I6, I7, I10);
    SUM1: OR_2 port map (I8, I9, I11);
    SUM2: OR_2 port map (I10, I11, Y);
end STRUCTURAL2;
```

Figure 4.11 4-to-1 multiplexer using simple logic gates.

The architecture that uses these components is called **STRUCTURAL3**. It is shown in Figure 4.13. As the **GATES** library is not implicitly visible to the working library, a library clause must be used. The synthesis environment must provide a mechanism for linking this logical library name to the physical library stored somewhere in the designer's directories. VHDL does not define this mechanism: this is left to the synthesis tool vendors. Note that there is no Use clause as this is required only to make particular packages or declarations within a package visible.

The architecture body illustrates another way to instantiate components in a design. Again, there are no component declaration statements in the architecture's declarative part. However, this time these declarations do not even exist in a package. Instead, the components are directly instantiated by specifying the name of the entity and architecture (in parentheses) in the component instantiation statement. The port list follows in the normal way.

```
-- 3 input AND gate
entity AND _3 is
    port (
            A, B, C : in BIT;
            D : out BIT
            );
end AND _3;
architecture DATAFLOW of AND _3 is
begin
    -- parentheses not required for expression with one type of logical
operator
    D<= A and B and C;
end DATAFLOW;
-- 4 input OR gate
entity OR_4 is
    port (
            A, B, C, D : in BIT;
            E : out BIT
            );
end OR_4;
architecture DATAFLOW of OR_4 is
begin
    E<= A or B or C or D;
end DATAFLOW;
```

Figure 4.12 Dataflow descriptions for 'optimum' gates. These are compiled and stored in a library called **GATES**. The inverter's description appears in Figure 4.9.

```
-- Make the GATES library visible to this entity and architecture.
library GATES;
-- Entity declaration shown in Figure 4.5.
-- 'optimum' gates
architecture STRUCTURAL3 of MUX4TO1 is
    signal I0,I1,I2,I3,I4,I5 : BIT;
begin
    -- Use direct instantiation of components (no declarations).
    -- Generate inverted control signals.
    INVERT0: entity INVERT(DATAFLOW) port map (S(0), I0);
    INVERT1: entity INVERT(DATAFLOW) port map (S(1), I1);
    -- Generate each product term.
    PROD0: entity AND_3(DATAFLOW) port map (I1, I0, A, I2);
    PROD1: entity AND_3(DATAFLOW) port map (I1, S(0), B, I3);
    PROD2: entity AND_3(DATAFLOW) port map (S(1), I0,  C, I4);
    PROD3: entity AND_3(DATAFLOW) port map (S(1), S(0), D, I5);
    -- Generate sum term.
    SUM: entity OR_4 port map (I2, I3, I4, I5, Y);
end STRUCTURAL3;
```

Figure 4.13 4-to-1 multiplexer using 'optimum' logic gates.

4.2.3 Synthesis and optimization

This section highlights the important aspects of the synthesis process for the multiplexer circuits designed in section 4.2.2. It is not a step-by-step guide to using synthesis tools but instead discusses how the differing designs affect the circuits that are generated by both logic synthesis and optimization. Again, each architectural style will be discussed in isolation.

Preliminaries

The synthesis process constructs a circuit from generic components. It is possible at this stage to target directly a technology vendor's library such that the generic circuit is automatically translated into the chosen library components. This can also be done manually as discussed in Chapter 2. Vendor libraries will often include macro blocks, technology-specific components or only certain types of logic gates that allow a design to be efficiently mapped into their own ASICs or programmable logic devices (PLDs). To illustrate the differences between circuits no particular technology needs to be targeted. To do so would introduce further complex, technology-dependent issues that would confuse rather than educate. These issues are, however, outlined in Chapter 2.

The synthesis tool will offer the designer a variety of compile-time options. As a typical example, the settings for the entity–architecture pair MUX4TO1–DATAFLOW1 could be as follows:

- *Directory and filename*: ./dataflow.vhdl
  Ensure that the VHDL source code is located in the current working directory or is pointed to by the pathname.
- *Type of input*: VHDL
  The synthesis tool will often accommodate C or Verilog HDL models.
- *Entity name*: MUX4TO1
  The tools are often case sensitive (even though VHDL itself is not).
- *Architecture name*: DATAFLOW1
  As only one filename has been specified it is assumed that the entity and architecture are located in the same file.
- *Save compiled VHDL Files*: no
  This option will remove all temporary and intermediate files used during the synthesis process. When compiling packages or other design units into libraries for use by other design units, the intermediate files will need to be stored. These are the physical files that are pointed to by the logical library / package names in Use clauses. They can be discarded when compiling this entity–architecture pair as it is not a library component.
- *Type of output*: ddb
  Some synthesizers may allow the output to be generated in a number of different formats. Usually a design database – a .netlist file – is required.
- *Share resources*: no
  This option is discussed in Chapter 6.
- *Logic partitioning*: no
  If logic partitioning is allowed, certain blocks in the design that are flagged by the designer are not optimized. This allows a design to compile faster. This option is generally used with arithmetic circuits and is discussed in Chapter 6.
- *Wired logic type*: OR
  Specifies the type of logic to be used when a signal has more than one driver or source. Options may be tristate (TRI), wired-And (AND) or wired-Or (OR). This option is irrelevant here and is discussed further in Chapter 7
- *flip flop type*: M
  Determines the type of flip flop that will be used in a sequential circuit. Options are generally for multiplexed data input type (M) or single input D-type (D).
- *Destination library*: generic
  The target technology library supplied by a vendor can be specified or the design can be compiled into a generic library, as is the case here.

An environmental parameter file may also be associated with the synthesis tool and could include information such as:

- the physical library path for the target technology and/or generic library;
- the operating characteristics of the circuit, including temperature and voltage.

A variety of statistics are generated by the synthesis tools for each circuit. When generic libraries are being used these data bear little similarity to the characteristics expected from a technology mapped design. However, they provide a clear basis for the comparison of the area and speed characteristics of the different circuits. The following statistics are presented for each circuit:

- *cells*: the number of unique components in the design;
- *transistors*: the number of transistors contained in all the (generic) cells;
- *area*: the area occupied by all the components, measured in area units – the effect of routing between components cannot be considered until the layout stage;
- *longest input to output delay*: the longest time taken for a change in any one of the inputs to become visible on any of the outputs (measured in nanoseconds).

In Chapter 2, the various optimization strategies were introduced. In this example, the circuits are only optimized for minimum area, illustrating what would always be the first step in the process.

Logic synthesis and optimization results

Synthesis of the dataflow architectures
The 4-to-1 multiplexer circuit produced by the synthesis of the architecture **DATAFLOW I** is shown in Figure 4.14. A combinational circuit has been constructed that uses only two-input And gates, Or gates and inverters. It implements every logical operation that is specified in the unconditional signal assignment statement of the architecture. For example, the signal S1 fans out into four branches, two of which generate the complement. The signals A and B are Anded with these inverted signals while C and D are each Anded with S1 directly. These four internal signals are then Anded at the next logic level with either S0 or its complement. Observe how the inputs are evaluated with the select lines in the sequential order that they appear in the expression – S1 first, then S0. The same ordering also occurs when each of the final four subexpressions are Ored together. One level of parentheses has been necessary to determine the precedence with which each element in the expression is evaluated. As a result, the logical Or operator is forced to have the lowest precedence.

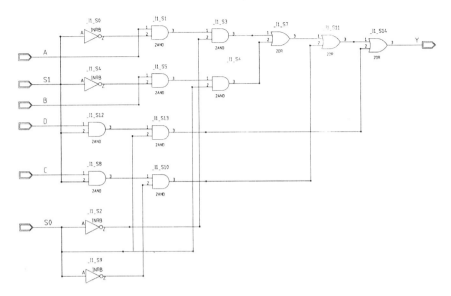

Figure 4.14 DATAFLOW1 synthesized circuit.

With a second level of parentheses, the structure of the circuit changes. Figure 4.15 shows the multiplexer circuit produced by the architecture DATAFLOW2. It looks similar to the circuit produced by DATAFLOW1, but the second level of parentheses has forced the pre-evaluation of the select logic – ...**not** S1 **and not** S0..., etc. – for each input. These four internal signals are then Anded with the appropriate input signal. The product operations are performed in the same order as before.

Despite these differences, the two circuits are still relatively similar when compared with the logic generated for DATAFLOW3, illustrated in Figure 4.16. This combinational circuit not only has an extra level of logic but also contains a more complex logic element in the form of a four-input Or gate. As will be seen many times throughout the examples in this text, language constructs such as the conditional signal assignment statement create the logic to implement the desired function in a predetermined way. For this particular construct, the logic evaluates the first condition and, if TRUE, assigns the associated expression to the output signal. Additionally, all the other conditions in subsequent When branches must be forced to evaluate FALSE and thereby avoid contention at the output. Thus, branch N can only evaluate TRUE if its particular condition is met and those in the previous N–1 branches were FALSE.

The branch conditions in the architecture DATAFLOW3 are evaluated by the first and second logic stages in the circuit. The next four levels provide the signal inhibition or selection logic. Only one of the four input signals is passed to the four-input Or gate, which finally generates Y. Hence, this

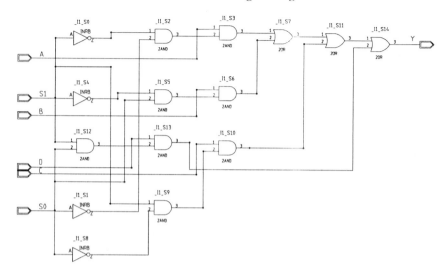

Figure 4.15 DATAFLOW2 synthesized circuit.

circuit has built-in protection against duplicated condition expressions in the signal assignment. Although the description of the circuit's function is clearer to follow in this architecture than in the previous two, its circuit is larger.

DATAFLOW4 produces a different style of circuit to the other three. As Figure 4.17 shows, this architectural description has been implemented with two-input Nor and And gates. Like the first two architectures it evaluates all four conditional expressions concurrently but then simply sends the four results to a four-input Or gate such as DATAFLOW3. This approach uses the concurrent hardware style of the unconditional assignment statements without the 'protection' against duplication. This is possible as the selected signal assignment statement checks for multiple occurrences of a condition at compile time, not in hardware. This makes the circuit smaller and faster

The statistical results for the logic synthesis of the four architectures are summarized in Table 4.4. Remember that generic synthesis libraries have been used and, as a result, the figures do not match those expected from vendor hardware technology libraries. The results confirm that the selected signal assignment architecture has produced the smallest circuit. The unconditional signal assignment architectures are next, but they are not identical owing to the change in the evaluation order in DATAFLOW2. Although the logic requires the same number of components, its interconnection pattern is different. This results in the signal path with the longest propagation delay being S0 to Y as opposed to S1 to Y. The synthesis tool has interpreted literally every specified logical operation. Hence, where only one

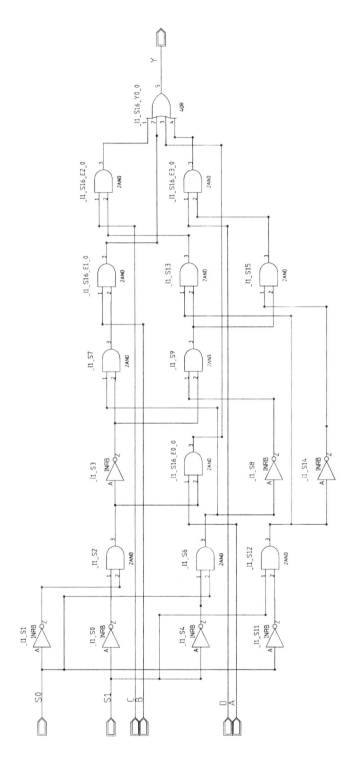

Figure 4.16 DATAFLOW3 synthesized circuit.

Combinational logic design

inverter is really necessary, two have been used. The optimization process has therefore been applied to all four circuits with the aim of removing this duplicated and any other redundant logic.

Optimization of the dataflow architectures
Table 4.4 also shows the statistical data for the optimized circuits. In each case, the level of optimization effort selected (low, medium or high) made no difference to the circuit that was produced.

The architectures **DATAFLOW1**, **DATAFLOW2** and **DATAFLOW4** produced an identical circuit after optimization. This is shown in Figure 4.18. The logic of the circuit has been minimized to the form shown in equation 4.4.

$$Y = (\overline{S0} + \overline{(B.\overline{S1} + D.S1)}).(\overline{S0 + (A.\overline{S1} + C.S1)}) \tag{4.4}$$

The circuit is constructed from only two inverters and three very simple macro blocks. The first two blocks, S152 and S153, evaluate the subexpressions containing S1 while the last block performs the sum and final product operations. The optimizer could have produced many different forms of this circuit so how, for example, did it choose to use a Nand gate at the output? This was probably because inverting gates are gener-

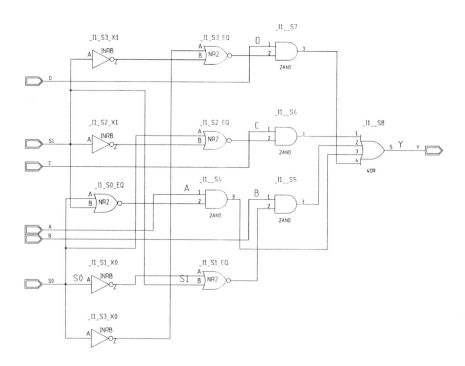

Figure 4.17 DATAFLOW4 synthesized circuit.

Table 4.4 Logic synthesis and optimization statistics for the dataflow style architecture

Parameters Temperature: 25.0 °C Voltage: 5.0 V	Synthesis Library: generic	Optimization Type: area Level: low
Circuit: MUX4TO1-DATAFLOW1		
Cells	15	5
Transistors	74	28
Area	9154.08	3344.76
Longest input to output delay	(S1 to Y) 3.171 ns	1.568 ns
Circuit : MUX4TO1-DATAFLOW2		
Cells	15	5
Transistors	74	28
Area	9154.08	3344.76
Longest input to output delay	(S0 to Y) 3.171 ns	1.568 ns
Circuit: MUX4TO1-DATAFLOW3		
Cells	19	10
Transistors	90	44
Area	11266.56	5809.32
Longest input to output delay	(S0 to Y) 4.619 ns	2.967 ns
Circuit: MUX4TO1-DATAFLOW4		
Cells	13	5
Transistors	58	28
Area	7393.68	3344.76
Longest input to output delay	(S1 to Y) 2.292	1.568

ally smaller than non-inverting ones. Even for straightforward circuits such as this one, it is not easy to determine the structure that will be generated by the logic optimization process and different optimizers are more than likely to produce different results. As will be seen in later examples, this can cause many problems in larger designs and calls for a greater appreciation from the designer of how and when parts of a circuit should or should not be optimized.

Note that the statistical parameters relating to the size of the circuit (cells, transistors and area) are independent of each other. In the optimized design, each macro block contains three logic elements but is counted as one cell. The number of transistors is determined by the type and number of each logic element. Finally, the area is not only calculated from the number of transistors, but is also affected by the complexity of individual components. For example, the function of the component S153 can be constructed using three independent gates rather than the simple macro, but the area would probably be larger.

The resulting circuit produced by the optimization process for DATAFLOW3 is illustrated in Figure 4.19. As the statistics for this circuit

Combinational logic design

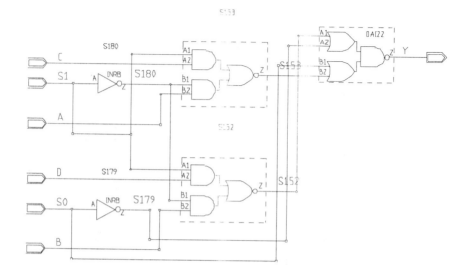

Figure 4.18 DATAFLOW1, DATAFLOW2 and DATAFLOW4 optimized circuit.

indicate, it is almost 74% larger than the other optimized circuits and must operate at almost half the frequency. This seems very strange for a circuit that is designed to perform exactly the same function! However, the conditional signal assignment statement imposes a different structure to the required logic function of the circuit by only allowing further When clauses to be evaluated if a condition is false. Such protection is not available in a concurrent assignment statement, which allows a 'free-form' style of logic to be constructed. The latter approach has its pros and cons. Circuits can be smaller but, as logic expressions become more complicated, the opportunity for both design errors increases.

Synthesis of the behavioural architectures

All four circuits constructed in the behavioural style have been synthesized and optimized using the same options and environmental parameters that were set for the dataflow architectures. In each case, the logic synthesis tool has generated a design with a two-level hierarchy. The top level for the CASE1 and IF_ELSE architectures is illustrated in Figure 4.20 (Figure 4.5 for the VHDL code). It seems that the synthesizer has generated a rather functionless circuit that could have easily been dispensed with. This is a result of the Process statement which will infer a hierarchical block when synthesized. If the architectures had contained any other concurrent statements outside of the process, their combinational logic would have been implemented at this top level. Note that the select bus S(1:0) has been divided into the two independent signals, S1 and S0.

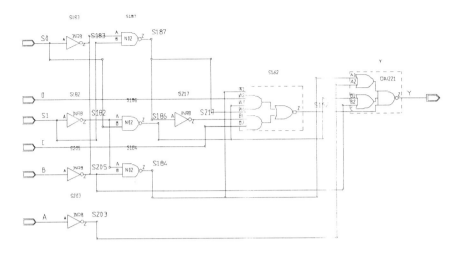

Figure 4.19 DATAFLOW3 optimized circuit.

Below this common block, the logic of **CASE1** and **IF_ELSE** is very different. The circuits are shown in Figures 4.21 and 4.22 respectively. It can be
seen that a **Case** statement will produce a circuit with fewer cells than when
an **If-Else** statement is used. More importantly, **CASE1** has only four stages
of logic compared with **IF_ELSE**'s seven. As the statistics in Table 4.5 illustrate, this gives the **CASE1** circuit half the propagation delay of **IF_ELSE**.
The beauty of the **Case** statement is that the condition attached to every
When clause is evaluated at the same time. Not only this, any possible
signal contention, as a result of multiple conditions evaluating true, is elimi-
nated at compile time and not in hardware. As a result, each **When** clause
can evaluate its expression independently of all the others. The one that
evaluates **TRUE** can then execute its sequence of statements safe in the knowledge that all other conditions have evaluated **FALSE**.

The only exception to the above rule is that the Others clause must wait
until all conditions have been evaluated to determine if the Case statement's default sequence is to be executed. The effect on the hardware of
using an Others clause can be variable, depending upon how many conditions are being tested and therefore how many input signal permutations
are being left to the default state. In CASE1, the final clause when '11' =>...
could have been replaced with when others =>..., but the logic this would
require to determine if the other three conditions were FALSE would be
more than that required to test the final condition, S="11". The Others clause
is most useful when only a few of the possible permutations have been
explicitly tested.

The **IF_ELSE** architecture unconditionally assigns Y to D if all other conditions are **FALSE**. This makes the logic of the circuit slightly simpler than

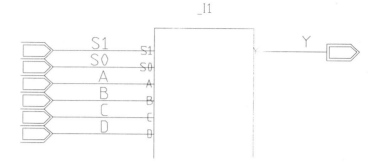

Figure 4.20 Top-level circuit for **CASE** and **IF_ELSE** architectures.

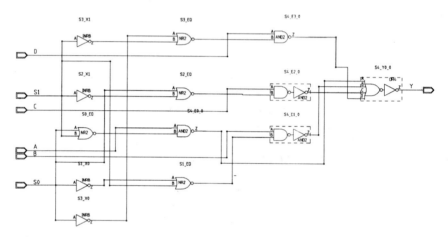

Figure 4.21 Second level of the synthesized **CASE1** circuit.

for the **CASE** architecture, which explicitly tests the final condition. Despite this, the **If-Else** statement still produces a larger circuit than the **Case** statement.

It is interesting to note that the circuit produced by the **IF_ELSE** architecture is very similar to the one generated by **DATAFLOW3**, shown in Figure 4.16. In this **If** statement, each branch is used only to assign a value to a signal – the same function performed by the conditional signal assignment statement in **DATAFLOW3**. However, being located inside a process, each branch of an **If** statement can perform any number of sequential operations, not just one signal assignment. It is therefore more flexible than its concurrent counterpart.

CASE2 and **CASE3** are the alternative behavioural designs that were shown in Figure 4.6. They are designed to illustrate how a variable can be

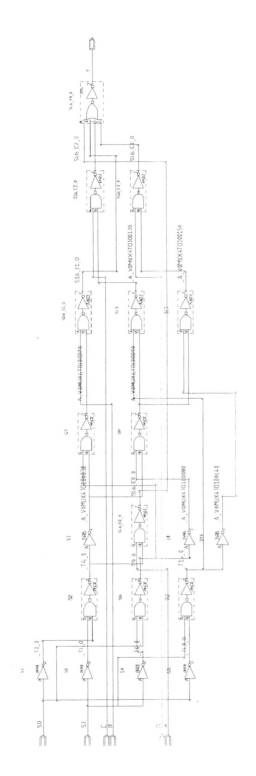

Figure 4.22 Second level of the synthesized IF_ELSE circuit.

Combinational logic design

Table 4.5 Logic synthesis and optimization statistics for the behavioural style architecture

Parameters Temperature: 25.0 °C Voltage: 5.0 V	Synthesis Library : generic	Optimization Type: area Level: low
Circuit: MUX4TO1-CASE1		
Cells	13	5
Transistors	58	28
Area	7393.68	3344.76
Longest input to output delay	(S0 to Y) 2.292	1.568 ns
Circuit: MUX4TO1-IF_ELSE		
Cells	19	10
Transistors	90	44
Area	11266.56	5809.32
Longest input to output delay	(S0 to Y) 4.619 NS	2.967 ns
Circuit: MUX4TO1-CASE2		
Cells	13	5
Transistors	58	28
Area	7393.68	3344.76
Longest input to output delay	(S0 to Y) 2.292 NS	1.561 ns
Circuit: MUX4TO1-CASE3		
Cells	13	5
Transistors	58	28
Area	7393.68	3344.76
Longest input to output delay	(S0 to Y) 2.292 NS	1.561 ns

used instead of a signal inside the Process statement. The top-level architectural blocks of the two synthesized circuits are compared in Figure 4.23. Their only difference is in the name of the output pin name on the block representing the process.

In CASE2, TEMP is a variable declared within the process. It can never be referenced outside and therefore it must be assigned to the output signal Y internally. The output pin name is therefore Y. In CASE3, however, TEMP is a signal that is declared in the architecture's declarative part. It can therefore be used inside and outside of any process within the architecture. It is assigned a value inside the Case statement and therefore becomes the output signal from the process, giving its name to the block output pin. Outside the process block it is assigned in a concurrent statement to Y – the output port declared in the entity. Inside the process block, the CASE2 and CASE3 circuits are virtually identical to the CASE1 circuit of Figure 4.21. The only difference for CASE3 is that the block output signal is called TEMP.

Optimization of the behavioural architectures
The logic optimizer has accomplished several tasks in these examples. It has:

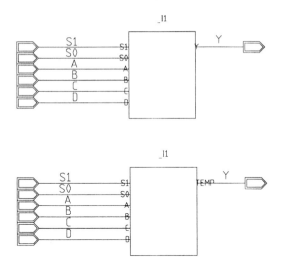

Figure 4.23 Top level circuits for the CASE2 and CASE3 architectures.

- stripped out the duplicated logic that was also present in the dataflow architectures;
- flattened the functionless hierarchy of each design;
- removed the unnecessary TEMP signal or variable from the Case constructs.

As the optimization statistics in Table 4.5 show, this has resulted in all three Case architectures producing the same circuit. This actually occupies approximately half the area and has half the propagation delay of the optimized IF_ELSE architecture.

The statistics for these two optimized circuits can be compared with those for the optimized dataflow architectures in Table 4.4. It can be seen that the statistics for the optimized architectures, CASE1, CASE2 and CASE3, are identical to those for DATAFLOW1, DATAFLOW2 and DATAFLOW4, as shown in Figure 4.18. In fact, all these circuits are identical. The same applies for the IF_ELSE architecture and the conditional signal assignment construct in DATAFLOW3, shown in Figure 4.19. The comparisons made earlier regarding the similarity of particular concurrent statements and sequential statements have therefore been validated.

These results are encouraging, as they show that the optimizer is successfully doing its job. This involves the removal of the VHDL elements that simplify the design process but complicate the hardware design. In most cases, this is what a designer wants the optimizer to do. But, as the examples in the structural architecture style will begin to illustrate, feeding a whole design into the logic optimizer and 'hoping for the best' is an approach that can be doomed to failure.

Synthesis of the structural architectures

The structural circuit descriptions illustrated how the multiplexer could be designed using a variety of simple components. The compilation of architectures containing components requires a little more thought than was applied in the other two design styles. In the structural style, a hierarchy physically exists in the description and is not just inferred by a Process statement, for example. When synthesizing the design, the process must begin at the bottom with all the components that do not contain any themselves. In the multiplexer, there are only two levels to the hierarchy. At the bottom, the components NAND_2, OR_4, etc. must first be compiled and stored in an appropriate library, making them available to other design units, such as STRUCTURAL1.

STRUCTURAL1 employs only Nand components. The logic synthesized circuit is shown in Figure 4.24. Each block represents a single instantiated component, as indicated by its label. The signals I0 to I5 are declared in the architecture for internal routing between components. The pin names allocated to each top-level block correspond to the ports declared in the components' entities. The external input and output signals are as declared in the Port statement of the entity MUX4TO1.

The contents of each discrete component block are shown in Figure 4.25a–c. These circuits are the result of dataflow style descriptions of two-, three- and four-input Nand gates. Just like the circuits generated for DATAFLOW1, etc., the circuits are very literal interpretations of the VHDL, and the order

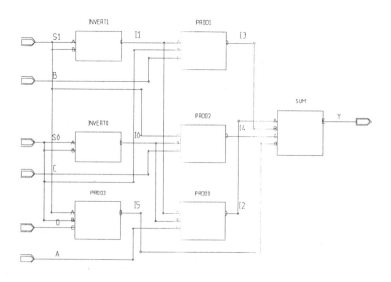

Figure 4.24 Top level of STRUCTURAL1 synthesized circuit.

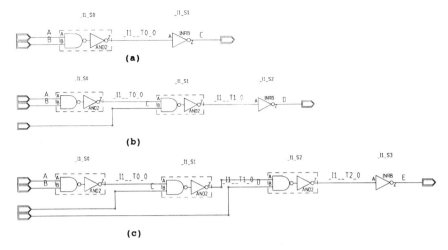

Figure 4.25 Two-, three- and four-input Nand gate components instantiated in STRUCTURAL1.

in which input pairs are evaluated depends on their order in the text of the description.

The other architectures, STRUCTURAL2 and STRUCTURAL3, produce similar circuits. In each case, all components instantiated in the architecture are represented at the top level of the design as hierarchical blocks. The top levels of each architecture are shown in Figures 4.26 and 4.27. The synthesized circuits for each component are shown in Figures 4.28 and 4.29. The statistics for all three architectures are shown in Table 4.6 and reveal some interesting results. In particular, the highest number of component instantiations are made in STRUCTURAL2 but, along with STRUC-TURAL3, this has the smallest area and lowest propagation delay. How is this possible?

Each component used in STRUCTURAL2 is composed of only a single two-input logic gate, whereas the other architectures employ slightly more complex combinational logic functions, with up to four inputs. At the top level of the design, more instances of simple components are required than of complex ones. The result is that the 'optimum' components in STRUC-TURAL3 and the 'minimum' components in STRUCTURAL2 just happen to generate the same circuit. When forced to use only one particular type of logic gate, such as the Nand in STRUCTURAL1, the resulting circuit is predictably larger.

Optimization of the structural architectures
The structural style architecture lends itself to a demonstration of the differing effects of optimizing a circuit by flattening or preserving its hierarchy. These techniques were discussed in section 2.4. The circuit statistics

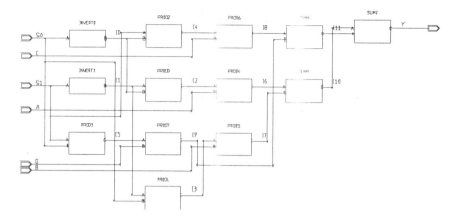

Figure 4.26 Top level of **STRUCTURAL2** synthesized circuit.

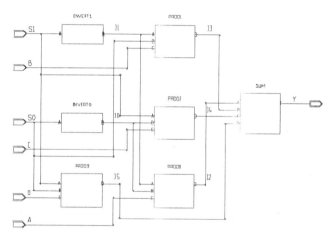

Figure 4.27 Top level of **STRUCTURAL3** synthesized circuit.

are shown alongside the original data in Table 4.6. It is clear that the smallest and fastest circuit is generated when the hierarchy is flattened. All three designs construct the same circuit, shown in Figure 4.30, when using this approach. This is almost identical to the optimized circuit generated by the dataflow and behavioural architectures (see Figure 4.18 for comparison), except that the logic of S0 is evaluated before that of S1. This occurs because the first input signal associated with the first component instantiation in the architecture is S0 in the inverter. This is the first signal the optimizer will encounter. The actual circuit has the same area and delay as five of the previously area-optimized multiplexer circuits.

On the other hand, preserving the hierarchy produces three different circuits. The optimization process is applied to every component individu-

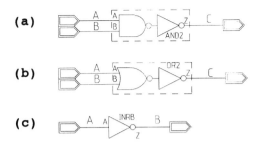

Figure 4.28 Components instantiated in STRUCTURAL2

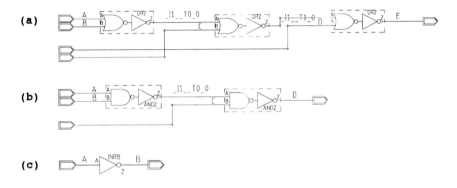

Figure 4.29 Components instantiated in STRUCTURAL3

ally. It therefore relies on each component having a capacity for optimization in the first place. The components used in STRUCTURAL2 are already in their simplest form and do not have this capability. The components in STRUCTURAL1 and STRUCTURAL3, however, can be optimized. The NAND_3 component, shown in Figure 4.25b, contains three And gates and an inverter. Optimization can directly replace all these elements with a single three-input Nand gate. The same applies to most of the other similar components.

As the statistics show, in all three circuits, flattening produces the better circuit. The hierarchical optimization process generates circuits with the number of logic cells equal to the number of instantiated components. Although STRUCTURAL1 and STRUCTURAL3 circuits each have seven cells, the former has a smaller area and a shorter propagation delay. This is because STRUCTURAL3 uses inverted Nand and Nor gates to create its And and Or logic, whereas STRUCTURAL1 uses Nand gates as its primary logic elements without any inverters. The reader should already be aware of the

Combinational logic design

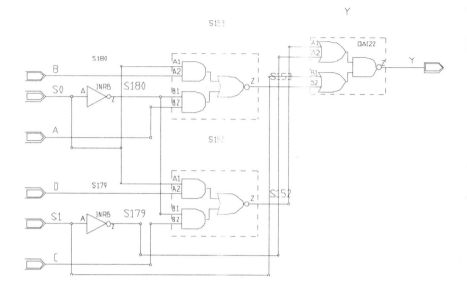

Figure 4.30 Structural style circuit flattened by optimization.

Table 4.6 Logic synthesis and optimization statistics for the structural style architectures

Parameters Temperature: 25.0 °C Voltage: 5.0 V	Synthesis Library: generic	Optimization Type: area Level: low	
		Flattened	Hierarchical
Circuit: MUX4TO1-DATAFLOW1			
Cells	20	5	7
Transistors	92	28	40
Area	11618.64	3344.76	4753.08
Longest input to output delay	(S1 to Y) 4.611 NS	1.568 ns	1.885 ns
Circuit: MUX4TO1-DATAFLOW2			
Cells	13	5	13
Transistors	70	28	70
Area	8449.92	3344.76	8449.92
Longest input to output delay	(S0 to Y) 2.304 NS	1.568 ns	2.304 ns
Circuit: MUX4TO1-DATAFLOW3			
Cells	13	5	7
Transistors	70	28	46
Area	8449.92	3344.76	5281
Longest input to output delay	(S0 to Y) 2.977 ns	1.568 ns	2.160 ns

fact that in both the CMOS and TTL logic families the Nand gate is the standard logic building block.

4.2.4 Multiplexer summary

Section 4.2 has illustrated a number of different ways in which a simple multiplexer can be designed in VHDL. Not all of these approaches are recommended for such a circuit, but the nature of the example has allowed a number of concurrent and sequential language statements to be introduced, not least various Signal assignment statements and the Case and If statements. For structural style architectures, a number of different techniques that allow components to be used in a design have also been demonstrated.

Differences in some of the synthesized circuits arose because of the language style and the constructs employed in each architecture. All the circuits broadly fell into one of three categories.

1 Those that evaluated every condition sequentially but still did not verify that only one of them was TRUE. These circuits contained Component instantiations or Concurrent signal assignments.
2 Those that evaluated each condition sequentially until one was found to be TRUE. These circuits contained If statements or Conditional signal assignments.
3 Those that evaluated all conditions concurrently but only allowed one to be valid. These circuits contained Case statements or Selective signal assignments.

The third category created the smallest circuit as the Case and Selective signal assignment statements ensure that conditions are not duplicated by checking at compile time and not in hardware. Logic is required by the second category of circuit to inhibit the evaluation of subsequent conditions once one is found to be valid.

After optimization, the largest circuits remain those in category 2, which require built-in inhibition logic. Circuits constructed using Case statements produce the smallest circuits, along with those in category 1. However, only a Case statement can guarantee full logic coverage for all possible states and without duplication.

Finally, circuits constructed in the structural style have components and therefore an explicit hierarchy. Flattening the hierarchy during optimization resulted in a circuit that was always smaller than when the hierarchy was preserved and only the components optimized.

4.3 BCD TO SEVEN-SEGMENT DISPLAY CONVERTER

This section presents a practical example of some of the language constructs introduced in section 4.2. The same approach is followed to the design process as used in the previous section except that only one solution is presented. The architectural style that is most appropriate for this particular example is used. As often occurs, this is a combination of two styles – behavioural and structural.

4.3 Functional overview

The circuit being design is a binary coded decimal (BCD) to seven-segment display converter. It receives as input an 8-bit pattern forming two BCD numbers and must generate an output to display the decimal equivalent value (00 to 99) on two seven-segment displays, as illustrated in Figure 4.31. For example, a bit pattern of 0001_1001 must display the decimal value '19'. The bit patterns 1110111 and 1100000 must therefore be output on LO_DRIVER and HI_DRIVER respectively. Segment 'a' is driven by the leftmost bit.

As with the example in section 4.2, all signals are of type BIT (two-value logic).

4.3.2 VHDL code design

This circuit contains two identical components. One drives the HI-DRIVER bus and the other the LO-DRIVER bus. The former decodes the upper 4 bits or **nibble** of the input and the latter the lower nibble. This suggests that a structural style would be an appropriate approach to adopt for a top-level architecture. This modular approach has the benefits of easier design and debugging, and also a simple expansion capability. Within each component, a straightforward transform is required from a 4-bit BCD to a seven-segment display bit pattern. Of the possible language constructs studied so far, it would seem that a Case statement is most appropriate for this task and therefore a behavioural style architecture is required.

The hierarchical structure of this design is illustrated in Figure 4.32. The top-level entity–architecture is CONVERTER-STRUCTURAL. This contains two instantiations of the component BCD2SSD-BEHAVIOURAL, uniquely labelled HIGH and LOW. The VHDL description of this component is shown in Figure 4.33 and the top-level design in Figure 4.34. The architecture body of the component contains a process with a Case statement. This assigns a bit pattern that forms the appropriate character, '0' through to '9', to the output bus DISPLAY. The Others clause is used to deal with the unused states of the input. The display lines are driven low if any illegal input pattern occurs.

In the entity CONVERTER, notice that the output buses HI_DRIVER and LO_DRIVER have been declared as type BIT_VECTOR with ascending ranges. This method is used because these buses carry a bit pattern and not a binary value. In such a situation it is not as important to stick with the convention that bit 0 is the least significant. In fact, in this case it is advan-

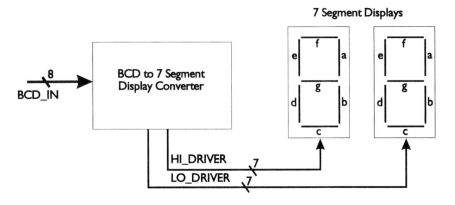

Figure 4.31 BCD to seven-segment display converter.

tageous to swap the notation around as LO_DRIVER(0) is now associated with segment 'a' on the display.

The component is compiled into the working directory in this case and therefore no explicit references to libraries are required in the **CONVERTER** design unit. Component instantiation has been made indirectly and so a component declaration is required. This has been placed in the declarative part of the architecture but could have appeared in a referenced package.

4.3.3 Synthesis and optimization

The approach that was applied to the structural style architecture examples in section 4.2 must also be used for this circuit. The component must be compiled first in the working directory followed by the entity–architecture pair, **CONVERTER–STRUCTURAL**. This top-level circuit will generate two instances of the component.

Preliminaries

This example again demonstrates the types of circuit produced by synthesis when using generic libraries. This circuit is optimized in its generic form. Some of the typical synthesis options for the component (top level) may be the following.

- *Directory and filename*: ./bin2ssd.vhdl (./converter)
- *Type of Input*: VHDL
- *Entity Name*: BCD2SSD (CONVERTER)
- *Architecture name*: BEHAVIOURAL (STRUCTURAL)
- *Destinationlibrary*: generic

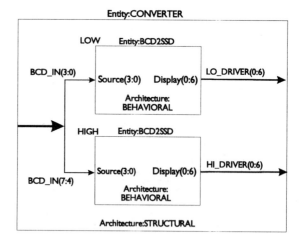

Figure 4.32 Structure of the VHDL code for the seven-segment driver.

```
entity BCD2SSD is
  port (
    SOURCE : in BIT_VECTOR(3 downto 0);
    DISPLAY : out BIT_VECTOR(0 to 6)
    );
end BCD2SSD;
architecture BEHAVIOURAL of BCD2SSD is
begin
  TRANSFORM: process(SOURCE)
  begin
    case SOURCE is
      when "0000" =>  DISPLAY<="1111110";
      when "0001" =>  DISPLAY<="1100000";
      when "0010" =>  DISPLAY<="1011010";
      when "0011" =>  DISPLAY<="1110011";
      when "0100" =>  DISPLAY<="1100101";
      when "0101" =>  DISPLAY<="0110111";
      when "0110" =>  DISPLAY<="0111111";
      when "0111" =>  DISPLAY<="1100010";
      when "1000" =>  DISPLAY<="1111111";
      when "1001" =>  DISPLAY<="1110111";
      when others =>  DISPLAY<="0000000";
    end case;
  end process TRANSFORM;
end BEHAVIOURAL;
```

Figure 4.33 VHDL description of the 4-bit to seven-segment display converter.

```
entity CONVERTER is
    port (
            BCD_IN : in BIT_VECTOR(7 downto 0);
            LO_DRIVER, HI_DRIVER : out BIT_VECTOR(0 to 6)
            );
end CONVERTER;
architecture STRUCTURAL of CONVERTER is
    component BCD2SSD
        port (
                SOURCE : in BIT_VECTOR(3 downto 0);
                DISPLAY : out BIT_VECTOR(0 to 6)
                );
    end component;
begin
    LOW : BCD2SSD port map (BCD_IN(3 downto 0), LO_DRIVER);
    HIGH : BCD2SSD port map (BCD_IN(7 downto 4), HI_DRIVER);
end STRUCTURAL;
```

Figure 4.34 Top-level structural description of the BCD to seven-segment display converter.

Synthesis results

The circuit that is generated by the synthesis tool has three layers. As demonstrated in the behavioural style 4-to-1 multiplexer circuits, a **Process** statement will create a process block that contains the circuit model. The top-level circuit is illustrated in Figure 4.35, while the component's block and circuit – the implementation of the **TRANSFORM** process – are shown in Figures 4.36 and 4.37. The top level of the design simply contains two instantiations of the component **BCD2SSD-BEHAVIOURAL**. The combinational logic inside the component reveals the full effect of the synthesizer's literal interpretations. Each input signal has multiple fan-outs to identical inverters. Other logic duplication also exists, but it is less obvious. Like the multiplexer example, the **Case** statement has evaluated all its branches concurrently. This results in a circuit with only four stages of logic, far fewer than the number that an **If-Else** statement would have created. The speed advantage of this tall circuit will be traded off against area during optimization. The statistics for one synthesized component are shown in Table 4.7.

Optimization results

The hierarchical design that synthesis has produced can be optimized in two ways:

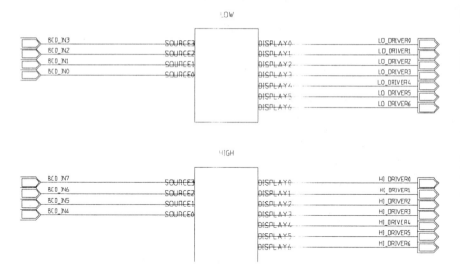

Figure 4.35 Converter top-level synthesized circuit.

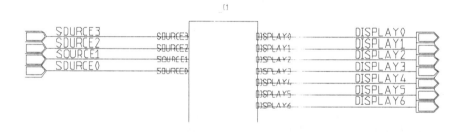

Figure 4.36 Component top level. The block represents the process **TRANSFORM**.

1 The component hierarchy could be preserved by optimizing each component individually and not optimizing at the top level at all.

2 The **CONVERTER-STRUCTURAL** entity–architecture pair could be optimized and the whole structure of the components flattened into one large circuit.

Which approach will lead to the smallest, fastest and (possibly) the 'best' circuit? Both techniques are attempted.

This circuit has three levels in its hierarchy, two of which are associated with the component. These two can be reduced to one without any major effect on the circuit as the process block is the only element within the component. Hence, the important structure of the design is still preserved if the component itself is flattened.

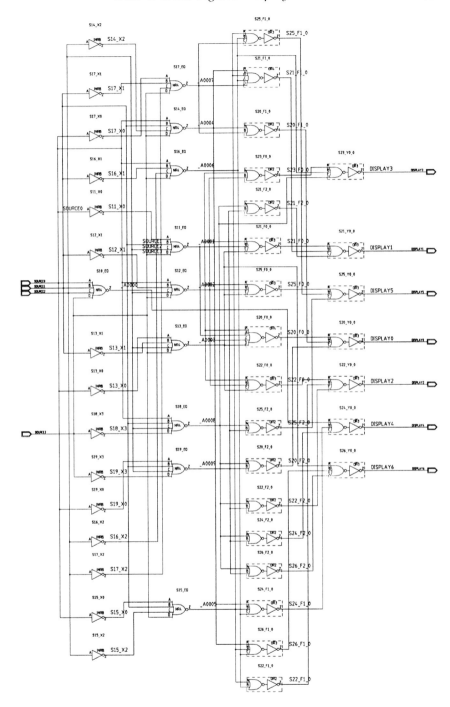

Figure 4.37 Implementation of the TRANSFORM process inside the component.

Table 4.7 Statistics for the unoptimized generic circuit

Parameters	Synthesis
Temperature: 25.0 °C	Library: generic
Voltage: 5.0 V	

Circuit: BCD2SSD-BEHAVIOURAL	
Cells	49
Transistors	288
Area	33975.72
Longest input to output delay	(SOURCE3-DISPLAY0) 6.733 ns

All three levels of optimization effort were attempted – low, medium and high. In the first instance, only the component is optimized (preserving the hierarchy) The statistics for each stage are detailed in Table 4.8. In this table the values have been doubled to account for two identical components. The component's final circuit, generated after a high level of effort, is shown in Figure 4.38. It would appear that all redundant logic has been removed and even the delay on the critical path has been reduced. The optimization process has more than halved the area of each component.

If this circuit is now compared with the flattened, optimized design, a good deal of similarities seem to exist. The circuit itself is not shown as it is very similar to a double version of Figure 4.38. Its statistics are however shown in Table 4.8.

The patterns in the statistical data for both circuits are very similar. The biggest single reduction in area occurs during the low-effort level of optimization, when the circuit area is cut by half. The medium-effort level produces no improvement and the high level only a slight one. After all three optimization processes the circuits are identical apart from the presence of the top hierarchical level (Figure 4.35) in the first circuit. As this level contains only two component blocks it is purely cosmetic and does not affect circuit area or delays.

As discussed in Chapter 2, each further level of optimization is characterized by an increase in CPU time as a more ambitious set of logic optimization operations are attempted. This example is typical of many, demonstrating that the lowest level of effort offers the best compromise between CPU time and circuit performance. For bigger circuits, the length of time to optimize at the higher levels can be many hours with only minor or no area improvements. An even worse scenario is that the area of a design could quite theoretically increase! A good philosophy to apply during area optimization is that when the design meets its constraints, leave it! Tackling the problems of meeting timing constraints when subject to area limitations is a topic discussed in Chapter 8.

Table 4.8 Optimized generic circuits with hierarchical and flattened structures

Parameters		Optimization	
Temperature : 25.0 °C	Type: area		
Voltage : 5.0 V	Level: low	medium	high
Circuit: CONVERTER-STRUCTURAL			
Component: BIN2SSD-BEHAVIOURAL (x2)*			
Cells (x2)*	52	52	42
Transistors (x2)*	264	264	228
Area (x2)*	32391.36	32391.36	27814.32
Longest input to output delay	6.107 ns	6.107 ns	5.298 ns
Circuit: CONVERTER-STRUCTURAL			
Component: all levels of the hierarchy			
Cells	52	52	42
Transistors	260	260	228
Area	32039.28	32039.28	27814.32
Longest input to output delay	5.579 ns	5.579 ns	5.298 ns

*The circuit contains two components.

4.3.4 Decoder summary

This example has illustrated how the VHDL constructs introduced in section 4.2 can be applied to the solution of a simple design problem. The decoder has demonstrated how the nature of the problem can influence the design style that is adopted for the architectural description of the circuit – to drive two identical displays two instances of one component were required. More instances of this same component can be declared if a larger converter needs to be constructed.

The two optimization processes demonstrated that flattening or preserving the hierarchy can produce similar results but this is circuit dependent, as will be shown in later examples. Employing a low level of effort generally provides the best trade-off between CPU time during optimization and circuit area.

4.4 DOS AND DON'TS

4.4.1 Array names

Don't declare array and scalar signals, variables or constants with the same prefix character(s). The array names will be modified by the synthesizer so that A(11) becomes A11, for example. It is simply best to avoid using simple names that end in a number when arrays have also been declared.

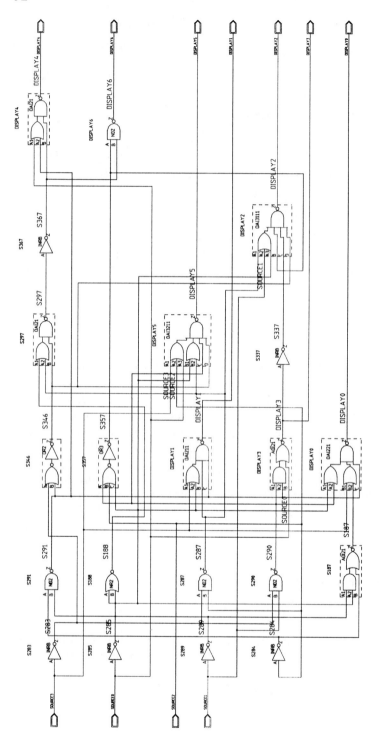

Figure 4.38 Optimized component BCD2SSD-BEHAVIOURAL.

4.4.2 Downto and to

Confusion and errors can be caused by mixing ascending and descending vector range indices. Do try to adhere to a particular convention.

The assignment of the binary value 10010 to two 5-bit vectors, A and B, declared in the two opposing fashions will occur as shown below.

signal A : BIT_VECTOR(4 **downto** 0);
signal B : BIT_VECTOR(0 **to** 4);

```
       MSB                    LSB
       A(4)  A(3)  A(2)  A(1)  A(0)
        1     0     0     1     0
       B(0)  B(1)  B(2)  B(3)  B(4)
```

Generally, a descending range should be used when the bit vector represents a binary value. The most significant bit is then always on the left. In some circumstances, such as to interface a signal to a schematic block, it may be necessary to use an ascending range. This can also be used when the bit vector represents a bit pattern and not a value. An ascending range was used in the converter in section 4.3 so that bit 0 represented segment 'a' on the display, and so on.

Using only one approach rather than both is not going to improve the area or speed of a synthesized circuit. What it will do however is ensure that simple, avoidable errors do not prolong the testing and debugging of a circuit.

4.4.3 Using operators with array types

When using logical or relational operators with operands of an array type, matching elements are compared by their indices not position. Hence, given the declarations and assignments:

signal A : BIT_VECTOR(0 **to** 4);
signal B : BIT_VECTOR(4 **downto** 0);
A <= "1000"
B <= "1100"
C <= A **or** B;

the result stored in C is 1011 not 1100 as A(0) is Ored with B(0) and so on. C must obviously have also been defined as a **BIT_VECTOR** and with an index range that matches that of the left-hand operator.

If the operands had been of different lengths, C would have been the length of the shortest with the index range direction of the left-hand operand.

4.4.4 Conditional branch evaluation order

In a conditional signal assignment, the evaluation of the condition in each When clause only occurs if the previous branch failed. The construct therefore generates a hardware multiplexer tree in which each branch contains a new two-input multiplexer. The same applies to a nested If-Else statement. This must be remembered when constructing the conditional expression for each branch. For example, consider the following statement:

```
Y<=  (A or B)    when (Q='0' and P="10") else
     B           when (P(I)='I')else
     (A and B)   when (Q='I' and P="II") else
     A;
```

If (Q='I' and P="II") is TRUE then B will always be assigned to Y instead of (A and B). It is clear in this case that the second condition is a subset of the third. The synthesizer will not complain about the logic of this statement. Do ensure that subexpressions are used carefully and are evaluated in the correct order.

4.4.5 The If statement

The free-form nature of the If-Else construct is open to abuse from the designer who is unaware of how it is translated into hardware. The safeguards built into the Case construct ensure full coverage of all input permutations. In the If statement, there is no validation of the conditions at compile time and if the designer so chooses, the conditional expression being evaluated can be different at each nested stage. Do use a default Else clause and don't mix highly different expressions within one If statement. Below is an illustration of a badly constructed if statement.

```
if A='0' then
    N<=3
elsif not P then
    M:=10
else
    N<=0;
end if;
```

Even though an Else clause has been included and the logic of the function may be correct, it is very difficult to follow. A better approach would be to use two separate If statements, one nested inside the other.

```
if A='0' then
    N<=3
else
    if not P then
```

```
      M:=10
   else
      N<=0;
   end if;
 end if;
```

Note that both If statements have default conditions that ensure that all possible input combinations are covered in the logic.

An If statement without an Else clause or one in which not all input permutations are tested is known as an incomplete If. The use of such a statement will move the design into the realms of sequential logic, a topic discussed in Chapter 5. It is sufficient here to discourage strongly the use of this language construct.

4.4.6 Using the unaffected keyword in signal assignments

The keyword Unaffected can be used in certain concurrent signal assignment statements to ensure that there is no change in the value of a signal given a particular condition. In this respect it is similar to the Null statement in sequential statement sections. Its use therefore implies the existence of feedback around the combinational logic. In some circuits this could possibly lead to hazardous conditions and so it should be used with caution.

4.4.7 Case versus If

The hardware consequences of using an If statement as opposed to a Case have been clearly demonstrated in this chapter. Do use a Case statement when possible as an alternative to an If statement. The following two examples have Case equivalents as shown. These will generate smaller and faster circuits. The signals S and P are of type BIT.

Example 1

```
process (S)                    process (S)
begin                          begin
   if S="00" then                 case S is
      Y<="001"                        when "10" => Y<="001";
   elsif S="01" then                  when "00" => Y<="101";
      Y<="101"                        when "01" => Y<="100";
   elsif S="10" then                  when others => Y<="111";
      Y<="100"                     end case;
   else                        end process;
      Y<="111"
   end if;
```

```
end process;
```

Example 2

```
process (S,P)                          process (S,P)
begin                                  variable SP : BIT_VECTOR(1 downto 0);
    if (S='1' and P='0') then          begin
        Y<="001"                       SP:= S & P -- concatenation
    elsif (S='0' and P='0') then           case SP is
        Y<="101"                           when "10" => Y<="001";
    elsif (S='0' and P='1') then           when "00" => Y<="101";
        Y<="100"                           when "01" => Y<="100";
    else                                   when others => Y<="111";
        Y<="111"                       end case;
    end if;                            end process;
end process;
```

In this second example, the two input bits S and P have been concatenated to form a 2-bit array signal. This is a useful technique, which avoids the construction of embedded Case statements. Circuits synthesized from code using embedded Case statements will be larger than their equivalent If or Case constructs. An example follows. This performs the same function as the second example above.

```
process (S,P)
begin
    case S is
        when '1' =>
            case P is
                when '0' => Y<="001";
                when others => Y<="111";
            end case;
        when '0' =>
            case P is
                when '0' => Y<="101";
                when '1' => Y<="100";
            end case;
        when others =>
            Y<="111";
    end case;
end process;
```

The Others clauses have been included here to allow for coverage of more general signal types, rather than just type BIT. These are discussed in Chapter 7.

In most cases the additional circuitry created by synthesis of **If** or nested **Case** statements can be removed by optimization. However, it is a waste of effort for the optimizer tools when the code could have been more compactly constructed in the first place. Don't nest **Case** statements and always choose a **Case** in preference to an **If** statement.

4.5 EXERCISES

The following exercises test the use of the language constructs that have been introduced in this chapter. They begin with simple problems and progress through to design exercises.

1 Determine which of the following logical signal and variable assignment operations are illegal.

   (a) Z <= A **and** B **or** C;
   (b) Z <= A **and** B **and** C;
   (c) Z := **not** A **nand** B **or** C;
   (d) Z <= A **nor** (B **nor** C);
   (e) Z := A **and not** B **and** C;

2 A 16-bit array signal, **OUTBUS**, of type BIT_VECTOR has been assigned the binary value 1000111000111110. The signal could have been declared with either of the following statements.

   (a) **signal** OUTBUS : BIT_VECTOR (15 **downto** 0);
   (b) **signal** OUTBUS : BIT_VECTOR (0 **to** 15);

   In case (a), what would be the values of the bits **OUTBUS(14)** and **OUTBUS(0)**?

   In case (b), what would be the values of the bits **OUTBUS(3)** and **OUTBUS(15)**?

3 Construct an entity that describes the top-level circuit shown below. Use signals of type BIT.

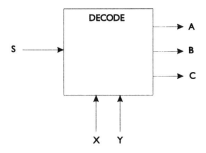

4 Design the architecture for the decoder in Exercise 3 (above) using a Case statement. Assume a straightforward decoding function, i.e. when X=0 and Y=0 then A=S, B=0 and C=0, etc.

5 Examine the function of the If statement shown below. Construct the *complete* architecture for a VHDL description that performs the same function with a Case statement. Assume all signals and variables are of type BIT or BIT_VECTOR.

```
if A='0' then
    S <= '1';
elsif B='1' then
    R := '0';
else
    S <= '0';
    R := '1';
end if;
```

6 Construct the entity declaration for the top-level block of the circuit shown below. Use the signal types encountered in this chapter.

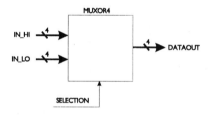

7 For the circuit shown below, construct the entity and architecture, labelling the instantiated components appropriately.

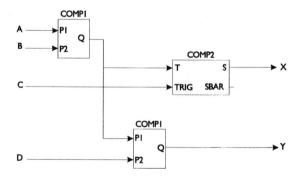

8 Design a two-input one-output multiplexer. The design should have two 1-bit inputs, A and B, a single-bit input selector line SEL and a one-bit output line C. $C = A$ if $SEL=1$, otherwise $C=B$.

9 Design a 4-bit comparator, which has two 4-bit unsigned (positive only) numbers A and B. There are three 1-bit output lines, GT, LE and EQ. GT is true if $A>B$. LE is true if $A<B$. EQ is true if $A=B$. Hint: If two of the outputs are calculated, then the third can be derived from the values of the other two (e.g. if A is not greater than B and not less than B, then it must be equal to B).

10 A router circuit is a key component in computing and telecommunication systems. The proposed circuit has four 1-bit inputs Ain, Bin, Cin, Din. There are four 1-bit outputs Wout, Xout, Yout and Zout. Each of the four inputs has associated with it its own 2-bit selector bits. Asel, Bsel, Csel and Dsel. If the selector lines associated with each input is 00 the relevant value on the in port is routed to Wout. If the selector lines are 01 the output is Xout, 10 means Yout is the destination and 11 means the destination is Zout. If the combination of selector lines is invalid (for example if two input signals are both routed to the same output), a 1-bit error signal is driven high and all four outputs are set to zero. Design such a circuit.

4.6 SUMMARY

This chapter has introduced the basic principles of designing combinational logic with VHDL. The reader should now be more familiar with the three architectural styles that are available and possess an idea of when and how they should be used in combinational logic design. Two operator types – logical and relational – have been explained, along with basic concepts such as naming objects and a range of signal assignment statements. A variety of methods for using components in a structural design have been explained through examples. The following points are worth repeating.

- The order of evaluation of concurrent signal assignment statements is not determined by their actual sequence in the architectural description.
- A process, any statement outside of a process and a component instantiation are all concurrent statements.
- Parentheses can be used to modify the order of evaluation of logical (and other) operators. The **Not** operator has the highest precedence of all logic operators. The rest are equal.
- A **Case** statement must provide coverage for all possible logical states of the signal under evaluation; an **If** statement does not.

- A Case statement will produce a smaller circuit when synthesized compared with an If statement.
- Variables are usually local to a process. Signals can be used to pass values or states out of a process.
- A conditional signal assignment statement is similar to an If-Else statement. The former is executed concurrently and the latter sequential (inside a process).
- A Selected signal assignment statement is similar to a Case statement. The former is executed concurrently and the latter sequentially (inside a process).
- A direct component instantiation does not require a Component declaration statement.

5

Sequential logic design

5.1 BASIC PRINCIPLES OF SEQUENTIAL DESIGN

5.1.1 Which architectural style?

The output of a sequential logic circuit is dependent not only on the values of the present inputs but also on the previous output. This implies that the state of the circuit must be stored internally. This is used to evaluate the new output state when a change in the input occurs or at regular fixed instances in time. A sequential circuit is therefore characterized by the notion of time. In VHDL it is possible to model both synchronous and asynchronous sequential circuits, but not using every design style. Dataflow-style circuits model the flow of data through a circuit in a continuous manner and possess no notion of time. This makes this style inappropriate for sequential logic design.

A structural style architecture can be used to model sequential circuits. The storage elements used in a sequential circuit are edge-triggered latches – flip flops – or level-triggered latches – transparent latches. These can be instantiated from libraries provided by the synthesis tool or technology vendor. This can limit the flexibility of an implementation as not all vendor technologies will support all types of storage elements. Elements such as flip flops with asynchronous preset and clear inputs, for example, may be omitted from a library.

An architecture in the behavioural style can infer sequential logic through the language constructs that it contains. These special statements are templates that tell the synthesizer to latch assigned signal values into storage elements. Statements associated with the template must describe the behaviour of a signal completely. Otherwise the synthesizer is left to determine this for itself – the behaviour is inferred. One problem with inferring elements is that it is difficult to determine exactly what type of storage elements will be used by the synthesizer.

5.1.2 Instantiating storage elements

By using a structural design style and explicitly declaring the elements (component instantiation) a number of benefits can be obtained over inferred elements in a behavioural style. In particular, the designer has full control of the implementation as:

- the function and type of each storage element will be precisely defined and not left to the synthesizer to select;
- the number of storage elements are determined by the number of instantiations;
- a technology can be targeted early in the design process to ensure good area and speed characteristics, the mapping of the circuit into a device is also a simpler and faster process.

5.1.3 Inferring storage elements

Inferring storage elements is a more complex process than instantiating them for three reasons.

1 To ensure that a storage element will be synthesized the sequential circuit's template must be correctly constructed.
2 The designer will not know the type of storage elements that will be used until the design has been synthesized.
3 It is very easy to generate unwanted storage elements if signals or variables are not treated correctly (section 5.1.4).

A flip flop or any storage element is inferred inside a **Process** statement. The template that indicates a sequential logic section is formed from either a **Wait** statement or an **If** statement.

Templates for sequential logic

A signal assignment inside the process shown below would infer to the synthesizer that the signal should be the output of a flip flop that is triggered by the signal **CLK**.

```
process
begin
    wait until (CLK'event and CLK='I');
    -- sequentially executed statements that
    -- should include a signal assignment
end process;
```

The **Wait** statement will be discussed in the next section, but this statement means '...when the signal CLK changes and that change is from '0' to '1', clock the function of the flip flop'. This is a rising edge-triggered flip flop. The Wait statement must always be the first in the process and all following statements are treated as part of the sequential logic section. A process with this structure is known as a **synchronous section**. Each statement in the section is modelled sequentially and executed each time the wait condition evaluates TRUE.

It is also possible to use an If statement to infer flip flops. In this case the statement

if (CLK'**event and** CLK='1') **then** ...

is interpreted as the template for the start of a synchronous section triggered by the signal CLK. It is therefore synthesized using flip flops. On the other hand,

if (CLK='1') **then** ...

could be a conditional statement in a combinational logic block or the template for a synchronous section. In this situation it is the type of assignments made within the statement that determines if a storage element will be inferred. Only a level-sensitive latch can be inferred from this statement, as discussed in section 5.4.4.

Some synthesis tools force the designer to specify which signals are clocks using signal attribute values. This helps to clarify ambiguous statements for both the synthesizer and the designer.

Unlike the Wait statement, the If does not have to be the first statement in the process. This allows both a synchronous section and combinational logic – a **combinational section** – to exist in the same process. The example below illustrates this (and assumes that all the signals have been declared in the surrounding architecture).

```
process (C, A, B, C, S)
begin
    -- combinational section
    A <= not B or S;
    if (B='1') then
        C <= S;
    end if;
    -- synchronous section
    if (CK'event and CK='0') then
        S <= A and C;
    end if;
end process;
```

Although only one synchronous section is shown here, any number of them could be described in a single process. In this example, the signal S is assigned a value *inside* the synchronous section whereas A and C are assigned values *outside* in the combinational part. The compiler will therefore automatically synthesize S as the output of a flip flop. A and C are inputs to the flip flops. An architecture that contains multiple synchronous sections must follow two key rules.

1 A signal assigned in one synchronous section cannot be assigned again in any other synchronous section, either in the same or a different **Process** statement.
2 A synchronous section cannot be nested inside another synchronous section.

These preclude a signal from being assigned on both a rising and falling clock edge or on different clocks.

This section has illustrated the important part that the **Process** statement plays in the design of sequential logic. In Chapter 4, its function in combinational design was also shown. It is good design practice to separate combinational and synchronous sections by using different processes. Box 5.1 offers some guidelines on how to use the process statement and illustrates its syntax.

What is inferred?

How does the designer know what the synthesizer will produce? This chapter should make clear the structure of inferred sequential logic circuits. However, the actual choice of storage element made by the synthesizer is one aspect of the design that is most often very difficult to predetermine. Although a library may have a range of different types available, including JK and SR types, the standard choice is usually a variant of the D-type. Any additional logic to implement a different type is then constructed on the front end of this.

This approach is adopted because the templates used to create synchronous sections do not provide any information regarding what type of logic is to be constructed. Analysing the contents of the sequential logic section to determine whether a JK type of flip flop was more appropriate than a D-type would increase the complexity of the synthesizer and the time taken to synthesize the circuit. Hence, using a simple type of flip flop with 'glue' logic is a trade off between synthesizer complexity/processing time and the area/speed characteristics of the design. As the user-defined JK flip flop in this chapter will illustrate, optimization can almost make up for the deficiencies of the synthesizer.

Conditions for inferring sequential logic

The examples in the following section illustrate clearly how to use **Wait** and **If** statement templates to infer sequential logic. The two general rules that will be demonstrated are as follows.

Box 5.1 How to use the Process statement

---

1 The process is a concurrent statement that allows the behaviour of small sections of a circuit to be modelled in a sequential fashion. It is considered to be the basic element of a behavioural architecture design style. The syntax for a process statement is:

```
process_statement ::=
  [label : ] process [ ( sensitivity_list ) ] [ is ]
        process_declarative_part
  begin
        process_statement_part
  end [ process ] [ label ];
```

2 The sensitivity list provides the trigger for the execution of the process. Any signal that could be read by the process, and hence affect its behaviour, should be included in the list (Box 5.8). However, when creating a synchronous circuit using a Wait statement (Box 5.6) the process is triggered solely by the signal in the Wait expression. A sensitivity list must not be used with a Wait statement.

3 The declarative header part contains a variety of declarations that may only be used within the process. These are the following:

   subprograms, types, subtypes, constants, variables, attributes.

   It may also contain subprogram bodies, Use clauses and attribute specifications.

4 The Process statement part may only contain sequential statements such as If-Else, Case and For Loop (Chapter 6). These are executed in the order in which they appear and the behaviour will be modelled with combinational and / or sequential (synchronous or asynchronous) logic.

5 Signal assignments inside a process do not take effect until the execution of the whole process is complete. See the Dos and Don'ts section.

6 Avoid multiple assignments to one signal within a process, unless each occurs on separate branches of a conditional statement, such as an If-Else or a Case.

7 Assignments to the same signal in different processes within one architecture are not allowed.

- A signal assigned inside a synchronous section is synthesized as a flip flop. If it is assigned inside a combinational section it will be synthesized as a signal on a wire.
- A variable that is used in the process *before* it is assigned a value is synthesized as a flip flop. Otherwise it is synthesized into a wire.

The same rules can be applied to arrays of signals and variables. In such a case, if one element of the array infers a storage element then so do all the elements. Level-sensitive latches are a special case discussed in section 5.4.4.

5.1.4 Initialization and undefined signals

For the designers of sequential circuits the initialization of storage elements to known states can be a thorny problem. If a combinational circuit is presented with valid input data, any undefined internal states will soon disappear and the circuit operation will be correct. However, the effect of undefined signals in the feedback loops of a sequential design pose a more serious problem. In the worst case the circuit could be locked in an unknown state.

When a design is simulated all signals and variables can be provided with initial values, avoiding any undefined states. However, this does not then guarantee that the circuit will function correctly when implemented in hardware, as each storage element will assume its own random state. Some mechanism must therefore be provided in the circuit to ensure that the storage elements are all correctly initialized or can be reset at any time to a known state.

The most common and definitely the best way of achieving this is to attach an external asynchronous or synchronous reset signal to each and every storage element in the circuit. When an external initialization signal is specified, the compiler will select an appropriate component from the library (if the elements are inferred). For example, if a flip flop is required to be initialized to '1', the selected element will possess an input for a preset signal. A variety of designs based on this form of initialization are given in the examples in this chapter.

The alternative way to deal with unknown or other such states is to 'catch' the good ones and 'throw out' the bad. This approach should be used with more caution as not all synthesis tools support the required constructs. The advantage of this method is that the VHDL can be constructed so that the simulation will accurately reflect the operation of the synthesized circuit design. With the previous method, this would only occur after the initialization signal had been issued. This topic is discussed in more detail in Chapter 8 when multivalued logic systems are introduced.

5.2 4-BIT SHIFT REGISTER

The 4-bit shift register is a simple circuit that suitably demonstrates the aspects of sequential logic design discussed above. Like the multiplexer example in Chapter 4, a number of different design methods are illustrated for the same circuit. Also, other new constructs are introduced as more complex flip flop functions are developed. Again, all the VHDL descriptions are synthesized and optimized to clearly demonstrate the difference between the alternative design approaches. The following language topics are covered in this chapter:

- array names
- array operators
- aggregate assignment
- positional and named association
- bit string literals
- interface lists
- concurrent statements:
     Process
     For-Generate
- Sequential statements:
     If (in a synchronous section)
     Wait (in a synchronous section)

5.2.1 Functional overview

The simplest form of the 4-bit shift register is shown in Figure 5.1. This version uses rising edge-triggered D-type flip flops that clock the input X through the register one stage per cycle. The output Z therefore appears after four clock pulses.

A number of variations on this design are presented using behavioural and structural VHDL. In particular, flip flops with synchronous and asynchronous preset and/or clear inputs are used, and falling-edge triggering is demonstrated. The output of the shift register, Z, is also inverted. This is done to add a little combinational logic to the design.

5.2.2 VHDL code design

Library selection

The basic requirements of this design are again met by the implicitly included libraries and packages. However, the std_logic_1164 package in the

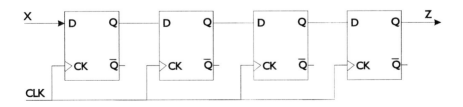

Figure 5.1 Simple 4-bit shift register.

IEEE library is also included in the architecture called **AGGREGATE** to make a clocking function visible. As in Chapter 4, signal and/or variable arrays could be declared to be of the unconstrained array type **BIT_VECTOR**. However, for 'standard' width vectors such as 4- or 8-bit it is often better to use predefined constrained array types. Tool vendors often provide a range of these or they can be user defined and stored in a package so that different designs may access them. The advantage of using predefined types in a package is the minimal amount of effort required to change the width of an array when a design needs to be modified. The user-defined package that has been created to store the new type declaration is shown in Figure 5.2. NEW_TYPE_DEFS only contains one declaration at present, a 4-bit array of type **BIT** called **HALF_REG**. This declaration could have been placed in an architecture's declarative part but would then only be visible to that particular design. Tool vendors will usually supply declarations very similar to this and as a rule these should be used when possible.

Another package called **NEW_PROCS** is shown in the figure. This contains a procedure called **PRESET_CLEAR**, which simply assigns a value, B, to the specified signal, A. The issues involved in declaring and calling procedures are not fully discussed until Chapter 7, but this procedure is shown here to allow it to be used in the initialization of the register. The std_logic_1164 package actually contains a procedure with the same name and operation but requires different types in its parameter list. This could have been used if type conversion (also discussed in Chapter 7) had been

```
package NEW_TYPE_DEFS is
  subtype HALF_REG is BIT_VECTOR(1 to 4);
end NEW_TYPE_DEFS;

package NEW_PROCS is
  procedure PRESET_CLEAR(signal A: out BIT_VECTOR; B:
BIT_VECTOR);
  attribute BUILT_IN : BOOLEAN;
  attribute BUILT_IN of PRESET_CLEAR: procedure is TRUE;
end NEW_PROCS;

package body NEW_PROCS is
  procedure PRESET_CLEAR(signal A: out BIT_VECTOR; B:
BIT_VECTOR) is
  begin
    A <= B;
  end;
end NEW_PROCS;
```

Figure 5.2 The user-defined packages NEW_TYPE_DEFS and NEW_PROCS.

performed on the parameters that were being passed to it. This new procedure does not replace the original one. Instead, either can be called, given
the correct set of parameters. This design method is known as **subprogram
overloading** and is discussed in Chapter 6. Attribute declarations and specifications and the BUILT_IN attribute in particular are discussed in Chap-
ter7.

The new packages are stored in a library called USER_UTILS. The
NEW_TYPE_DEFS package must be made visible to each design unit by
using Library and Use clauses. The procedure PRESET_CLEAR is only used
in the architecture SYNC_WAIT.

The entity declaration

The simple shift register design was illustrated in Figure 5.1. Generally,
each flip flop will also possess preset and/or clear inputs to allow the circuit to be initialized to a known state. In both behavioural and structural
design styles this can be obtained from a single INIT signal. The entity for
this circuit, called FOUR_SHIFT is shown in Figure 5.3. Note that a convention has been adopted for grouping signals – input first followed by control and finally output. Included in this file above the design unit are Use
clauses for the NEW_TYPE_DEFS package, stored in a user-defined library
called USER_UTILS.

The architecture body

Behavioural design style
The full VHDL description of this design is given in Figure 5.4. The architecture declares a signal called REG of type HALF_REG. The input X is
stored in REG(1) and the output Z is derived from REG(4). The architecture
contains two concurrent statements – the process and the unconditional

```
library USER_UTILS;
use USER_UTILS.NEW_TYPE_DEFS.all;
entity FOUR_SHIFT is
    port (
            X : in BIT;
            CLK : in BIT;
            Z : out BIT );
end FOUR_SHIFT;
```

Figure 5.3 Entity FOUR_SHIFT and the libraries/packages required for the
general form of the shift register.

signal assignment. The process will create its own sequential block, the output of which is the signal REG(4). This is inverted and the final output signal, Z, is generated.

The **Process** statement contains the description of the sequential logic that is implemented with flip flops by the synthesizer. A discussion on a variety of aspects related to processes was given earlier in Box 5.1. This process uses an If statement with a clocking expression to infer that the signal REG, which is assigned a value within the synchronous section, should be synthesized as a flip flop. Only when a signal is assigned under a clocking condition will it be synthesized as a flip flop. This If statement waits for a rising edge on the signal CLK. The new pattern to be stored in REG is then generated by concatenating X with the three least significant bits of REG. A portion of an array such as REG(1 **to** 3) is known as a **slice**. Addressing array elements using this method and others are discussed in Box 5.2. This is one way to perform bit-shifts in an array. However, it is also possible to use purposely designed shift and rotation operators that allow shift-by-N operations. These are discussed in Box 5.3 along with the concatenation operator that was encountered several times in Chapter 4.

Section 5.1.4 highlighted how important it was to ensure that sequential circuits are correctly initialized. The **SIMPLE** architecture does not provide any means of initializing the register to a known state and is therefore a poor design. Figure 5.5 shows an architecture called **STRING_ASSIGN**, which describes a circuit with synchronous initialization. The entity now contains the external initialization signal INIT, which has also been added to the process sensitivity list. Inside the process there is still only one If statement with a conditional clock expression – all conditional expressions nested inside the synchronous section are not considered to be clocking expressions. When a rising edge is detected, one of the following actions will occur:

```
-- libraries, packages and entity as shown in Figure 5.3
architecture SIMPLE of FOUR_SHIFT is
signal REG : HALF_REG;
begin
    process (X, CLK)
    begin
        if (CLK'event and CLK='1') then
            REG <= X & REG(1 to 3);
        end if;
    end process;
    Z <= not REG(4);
end SIMPLE;
```

Figure 5.4 SIMPLE architecture of a 4-bit shift register.

Box 5.2 How to manipulate array types

Array types can be used in the declaration of signals, variables and constants. How to make assignments to objects of this type is discussed later in this chapter in Box 5.4. As a precursor to that, this section looks at the ways in which the elements of composite types can be manipulated for use as targets or sources in assignment statements. Consider the types, subtypes and signals declared below.

```
type ROW is array(7 downto 0) of BIT;
type MEMORY_A is array(0 to 16, 7 downto 0) of BIT;
type MEMORY_B is array(0 to 16) of ROW;
subtype CTRLBUS is BIT_VECTOR(0 to 3);
subtype DATABUS is BIT_VECTOR(7 downto 0);
signal A : MEMORY_A;
signal B : MEMORY_B;
signal C : CTRLBUS;
signal D : DATABUS;
signal X : BIT;
signal Y : BIT_VECTOR(0 to 5);
```

The following illustrate how *indexed names* can be used to select elements of one- and two-dimensional arrays (1D and 2D).

```
X <= C(2);          --scalar <=scalar element of 1D
Y(1) <= A(0,0);     --scalar element of 1D <= scalar element of 2D
A(4,7) <= B(4)(7);  --scalar element of 2D <= scalar element of 1Dx1D
D <= B(2);          --1D array <=1D array element of 1Dx1D
```

Sliced names allow a continuous range or slice of elements from a one-dimensional array to be selected, as the following examples show.

```
C <= Y(0 to 3);              --1D array <=slice of 1D array
D(3 downto 1) <= Z(6 downto 4);  --slice of 1D array <= slice of 1D array
Y(1) <= Y(5 to 5);           --element of 1D array <= slice of 1D array
```

The following are examples of illegal index and slice names.

```
A(10,7 downto 0) <= B(10);    -- slice of a 2D array is not allowed
Y(1 to 3) <= Z(7 downto 5);   -- range directions must match on each side
Y(5 downto 2) <= Z(6 downto 3); -- values must be in the declared range
X <= B(1);                    -- type mismatch, BIT <= BIT_VECTOR
X <= C(2 to 2);               -- scalar <= array slice containing one element
```

- if INIT is not active (= '1') then the shift operation is executed;
- when INIT is low, initialization occurs by storing the bit pattern 1000 in REG.

Box 5.3 How to use array operators

The concatenation operator (&)

One-dimensional arrays and elements of the array's type can be combined with the concatenation operator. In the STANDARD package the operator is defined for the types BIT_VECTOR and STRING (which is not supported for synthesis). The target of an assignment must be an array type. The following examples use the declarations from Box 5.2 to illustrate legal and illegal concatenations.

```
Y <= C & "10";                    -- array <= array & array
Y <= C(0 to 2) & D(6 downto 4);   -- array <= array slice & array slice
Y(1 to 3) <= C(0 to 2) & X;       -- array slice <= array slice & element
Y <= C & "1110"                   -- source longer than target
X <= C(1 to 1) & ""               -- target must be a one-dimensional array type
```

Shift operators

The STANDARD package declares shift operators for one-dimensional arrays with elements of type BIT and BOOLEAN. The following operators are available:

sll	shift left logical	**srl**	shift right logical
sla	shift left arithmetic	**sra**	shift right arithmetic
rol	rotate left logical	**ror**	rotate right logical

Examples of each are shown below given that D initially contains 10010111.

```
D <= D sll 4;      -- D is "01110000", all bits shift and fill with '0'
D <= D srl -1;     -- D is "00101110", the same as 'D sll 1'
D <= D sla 2;      -- D is "11011100", sign bit unchanged, fill from left with '0'
D <= D sra 4;      -- D is "11111100", sign bit fills from the right
D <= D rol 3;      -- D is "10111100"
```

REG is assigned its new value through a bit-string assignment statement. This form of assignment is possible when each element of the string has a discrete encoded enumeration value, such as the type BIT. The more general and preferred method of string assignment is to use an **aggregate expression**. This is demonstrated in the next version of the circuit shown in Figure 5.6 and discussed in Box 5.4. Different ways of expressing strings of type BIT are illustrated in Box 5.5.

Figure 5.6 shows the entity for the shift register and an architecture body called AGGREGATE. The port list shown in the entity differs from the one in STRING_ASSIGN as CLK is now of type std_ulogic. This is declared in the std_logic_1164 package and discussed further in Chapter 7. Here it is used

```
library USER_UTILS;
use USER_UTILS.NEW_TYPE_DEFS.all;
entity FOUR_SHIFT is
   port (
         X : in BIT;
         CLK, INIT : in BIT;
         Z : out BIT
      );
end FOUR_SHIFT;

-- string assignment to an array
architecture STRING_ASSIGN of FOUR_SHIFT is
signal REG : HALF_REG;
begin
   process (X, CLK, INIT)
   begin
      -- different clock expression template to SIMPLE architecture
      if (not CLK'stable and CLK='I') then
         if INIT='I' then
            REG <= X & REG(I to 3);
         else
            REG <= "I000";
         end if;
      end if;
   end process;
   Z <= not REG(4);
end STRING_ASSIGN;
```

Figure 5.5 Inferring a flip flop with a synchronous preset and clear.

because the clock detection function rising_edge requires as its parameter a signal that is of this type. There are three possible solutions to this type of mismatch problem.

- Declare a new function of the same name that accepts the required signal type. This approach was used in the overloading of the PRESET_CLEAR procedure (Figure 5.2).
- Perform a type conversion on the signal so that one of the correct type is sent.
- Declare the signal to be of the correct type in the first place.

As CLK is only read by the process in the clock expression the last approach has been demonstrated. Using multivalued logic types such as std_ulogic makes the design process a little more complex and is discussed fully in Chapter 7.

```
library IEEE, USER_UTILS;
use IEEE.STD_LOGIC_1164.all;
use USER_UTILS.NEW_TYPE_DEFS.all;
entity FOUR_SHIFT is
    port  (
            X : in BIT;
            CLK : in std_ulogic;
            INIT : in BIT;
            Z : out BIT
            );
end FOUR_SHIFT;

-- aggregate assignment with positional association
architecture AGGREGATE of FOUR_SHIFT is
signal REG : HALF_REG;
begin
    process (X, CLK, INIT)
    begin
        -- clock detection function declared in std_logic_1164 package
        if rising_edge(CLK) then
            if (INIT='1') then
                REG <= X & REG(1 to 3);
            else
                REG <= ('1', others =>'0');
            end if;
        end if;
    end process;
    Z <= not REG(4);
end AGGREGATE;
```

Figure 5.6 Alternative architecture for the shift register with synchronous initialization and aggregate array assignment.

The initialization statement in the architecture uses the Others keyword to assign '0' to all bits in the register apart from the first. In the general case, this keyword will assign a value to any remaining unassigned elements of the array. This is a useful shortcut when the length of an array type definition is liable to be changed *before* the design is compiled. The clause can only be used with constrained composite types (the Dos and Don'ts section). In a simulation environment, the technique allows the definition of a type to be delayed until some other parameter is known. This is an **incomplete type definition**. This flexibility cannot be exploited when synthesizing, owing to the obvious restriction on creating hardware that can reconfigure itself depending on the value of a particular input! The main advantage of using this construct when synthesizing is that, if a type definition changes length, both the target and source of the assignment are automatically modified.

Box 5.4 How to make array assignments

Box 5.2 demonstrated how elements and slices of arrays can be used in assignment statements. The concatenation operator and shift operators can also be used in the same way as was illustrated in Box 5.3. Any of these manipulation techniques can be combined with others to make a more complex expression. For example, using the declarations from Box 5.2, arrays could be constructed as follows:

```
Y <= X & "01111";
Y <= '0' & X & (C(2 to 3) and Y(0 to 1) & "000");
```

Note the use of parenthesis around the logical function. Without these around the And operation the expression will be incorrectly evaluated. Both these expressions contain strings of characters enclosed in double quotes that are used to represent bit patterns. These examples illustrate binary representations but octal and hexadecimal formats may also be used (see Box 5.5). Using the concatenation operator to construct array expressions is only one approach. It is also possible to use aggregate expressions that take the form:

```
Y <= ( X,'0','1','1','1','1' );
Y <= ( '0', X, C(2) and Y(0), C(3) and Y(1),'0','0' );
```

In an **aggregate** assignment expression, each element of the array is individually assigned its value using a list format. The array could be composed of BIT, INTEGER or any other synthesizable elements. In the two examples above, the use of **positional association** is demonstrated – the physical position of an expression in the list determines which element in the array is assigned the value. The two examples below illustrate named association.

```
Y <= (0=>X, 1=>'0', others=>'1');
Y <= (0 | 4 to 5=>'0', 3=>C(2) and Y(0), 2=> C(3) and Y(1), 1=>X);
```

Named association allows the elements' expressions to appear in any order in the list and allows groups of elements with the same value to be grouped using:

- the set symbol, '|';
- a continuous range expression, with the keywords To or Downto;
- the Others keyword, for any as yet undefined elements.

An integer must be used to denote the array position of an element and the same position index must not be repeated. Remember that whichever assignment method is used, each element's expression must evaluate to the same type as the target. It is possible to mix positional and named association in the same list. However, position-associated elements must precede all name-associated elements, limiting the usefulness of this approach.

Named association can be used in other types of list (Box 5.8).

Box 5.5 How to use bit-string literals in arrays

An aggregate expression constructs an array from scalar types. A string expression may use objects of a composite type or explicitly specify the value to be stored in the array. The common way to express a binary value is to use a **string literal**, viz.:

    D <= "10011011";

String literals can generally only be used in synthesis if each character of the string represents a discrete enumeration literal and hence a single bit encoded value. Hence, an array of type **BOOLEAN** could be formed using aggregates but not strings as illustrated below.

    constant K : array(1 to 3) of BOOLEAN := ('TRUE', others=>'FALSE');-- legal
    constant L : array(1 to 3) of BOOLEAN := "TRUEFALSEFALSE";   -- illegal

The assignment made to D above is a special form of string literal as it represents a binary pattern. It is therefore equivalent to the binary form of a bit-string literal. A **bit-string literal** is a string of base type BIT that may contain **extended digits** ('0' to '9' and 'A' to 'F') and the underscore character. It is preceded by the **base specifier** character (B, X or O) representing a value in either binary, hexadecimal or octal form. The assignments below are all equivalent to string literal assignment to D above.

    D <= B"1001_1011";   -- the underscore is optional
    D <= X"96";
    D <= O"233";

If no base is specified then binary is assumed. This means that only the characters '0', '1' or '_' can appear in the BIT_VECTOR string unless a specific base is indicated. Similarly, when specifying the value in octal form only the extended digits '0' to '7' can appear in the string. Note that a bit string is not equatable to an integer value. For example, B "10011011" is not equatable to the integer value 155. This is because a string does not have a discrete value associated with it and therefore cannot be treated like an integer. See the Dos and Don'ts section for further elaboration on this point. It is still possible to use operands of different types, however, through the use of type conversion functions. These and other functions are discussed in Chapter 7.

The form of aggregate assignment shown in this example associates elements of the array with particular values by the position of the values in the list. It is also possible to form this association by name. Positional and named association can be used in many different types of list, as later examples will show, but they are introduced in Box 5.4 in the context of array aggregates.

The clock expressions are different in each of the three architectures above. The SIMPLE architecture uses the event attribute already seen in section 5.1.2. In the architecture STRING_ASSIGN, the use of another attribute called Stable is demonstrated. The expression must again test to see whether a change has occurred and therefore determines whether the clock signal is *not* stable. In the AGGREGATE architecture a predefined function called rising_edge() is used. This is defined in the std_logic_1164 package, and synthesis tool vendors will usually provide their own versions. Attributes are discussed fully in Chapter 7.

Flip flops that trigger on a falling edge can be inferred by modifying the clock expression to '...and CLK='0'". A predefined function called falling_edge() also exists. Note that within the same architecture body a single clock signal cannot be used to trigger both rising and falling edge flip flops, even if the synchronous sections are located in independent processes.

More details on inferring flip flops with If statements are given in Box 5.6.

The next set of examples demonstrates how synchronous and combinational sections can coexist in a process. Again an option for initializing the circuit is provided but this time with an **asynchronous** external signal.

Section 5.1.2 illustrated how a block of combinational logic can form part of a process that infers a flip flop. This block does not need to be associated with the logic in the synchronous section(s) of the process, although if it is not then it should usually be placed in a separate process. Generally, the combinational section should be used to implement any logic that does not need to be located within the synchronous section. Using this approach, synchronous blocks with minimum clock-to-output propagation delays should result.

Two architectures with asynchronous initialization are shown in Figure 5.7. ASYNC_IF1 illustrates an architecture with completely separate combinational and synchronous sections. Alternatively ASYNC_IF2 shows how the synchronous section can hang off the combinational section. In this example, note that the clock expression is on the *last* branch of the If statement. It would be an error if this statement was followed by another Elsif or an Else statement.

In ASYNC_IF1, if INIT is low then the register will be initialized to 1000, independent of the state of the clock. It will be held in this state while INIT remains low, regardless of the number of clock pulses received. The synchronous and asynchronous sections should not drive REG at the same time. To ensure this does not happen, a conditional statement within the synchronous section determines the state of INIT before the bits in the register are shifted. This will require additional logic for its implementation.

ASYNC_IF2 demonstrates how the two sections can be combined in one If statement. The circuit operation is exactly the same in this architecture as

Box 5.6 How to create a synchronous section with an If statement

1 An If statement can be used as a template to infer a flip flop within a process, creating a circuit with synchronous behaviour. Apply the following syntax:

```
process (sensitivity_list)
begin
    ...
    if clock_expression then
            --sequential statements, including signal assignments
    end if;
end process;
```

2 The sensitivity list is mandatory. It must contain all signals read by the process.

3 The clock expression must be one of the following:

```
(clock_signal'event and clock_signal=clock_value)
(not clock_signal'stable and clock_signal=clock_value)
```

Alternatively, tool vendors usually supply clocking functions. The std_logic_1164 package contains the following:

```
rising_edge(clock_signal)
falling_edge(clock_signal)
```

4 The clock signal is an identifier for a scalar signal or single element of an array (with a fixed index). In some synthesizers it is necessary to indicate that this signal is a clock by using an attribute (Chapter 7).

5 The clock value is '1' for rising edge or '0' for falling edge detection.

6 The use of the Event and Stable attributes is limited to the If and Wait constructs when synthesizing a design. See Chapter 7 for more details on attributes.

7 The sequential statements inside the If *must* assign a value to a signal or to a previously undefined variable for the synthesizer to infer a flip flop and hence create a registered output. Otherwise, purely combinational logic will be generated.

8 The If statement can appear anywhere in the process but:

   • it must not have an Else or Elsif clause following it, and
   • it must not be nested inside another If.

   However, it can be the final Elsif branch of a combined combinational and sequential circuit inferring If statement. Examples are given in this section.

9 Any number of synchronous sections may appear in a single process and any number of processes within the architecture may contain synchronous sections.

10 A single clock signal cannot drive both rising and falling edge flip flops within the same architecture.

```
--libraries, packages and entity as shown in Figure 5.5
```

-- Independent asynchronous initialization

```
architecture ASYNC_IF1 of FOUR_SHIFT is
signal REG : HALF_REG;
begin
  process (X, CLK, INIT)
  begin
    if (INIT='0') then
            REG <= ('1', others=>'0');
    end if;
    if (CLK'event and CLK='1') then
            if (INIT='1') then
                    REG <= X & REG(1 to 3);
            end if;
    end if;
  end process;
  Z <= not REG(4);
end ASYNC_IF1;
```

-- Asynchronous initialization combined with the synchronous section

```
architecture ASYNC_IF2 of FOUR_SHIFT is
signal REG : HALF_REG;
begin
  process (X, CLK, INIT)
  begin
    if (INIT='0') then
            REG <= ('1','0','0','0');
    elsif (CLK'event and CLK='1') then
            REG <= X & REG(1 to 3);
    end if;
  end process;
  Z <= not REG(4);
end ASYNC_IF2;
```

Figure 5.7 Shift register architectures with asynchronous initialization.

in ASYNC_IF1. This second design should produce a smaller circuit than ASYNC_IF1 when first synthesized. After optimization, however, both circuits should be identical.

In all the above cases, the sensitivity list and the style used to construct the If statements – the templates – enables the synthesizer to determine which sections of the code are synchronous and which are combinational.

If a Wait statement is used instead of an If, the sensitivity list is not used and the process is fully synchronous.

Figure 5.8 illustrates an architecture called SYNC_WAIT containing a synchronous process formed from a Wait statement. This version of the shift register uses the user-defined procedure for presetting and clearing REG called PRESET_CLEAR. The additional Use clause is specific in making that particular procedure in the package NEW_PROCS visible. There is no sensitivity list and therefore only the clock signal, CLK, within the Wait can trigger the process.

If the initialization signal, INIT, is high the normal shift operation occurs. The register will be initialized if INIT goes low. These operations inside a process are synchronous when a Wait statement is used with the clock expression. The trigger in this case is a rising clock edge. See Box 5.7 for more details on this construct.

Structural design style

A structural style description of the 4-bit shift register can be constructed in two ways. The most straightforward approach is to instantiate storage elements with the required functionality from vendor-supplied libraries. If the required type of flip flop is not available, the designer may choose to construct a component with the required function using a behavioural description. Ultimately, this approach will still produce a circuit with storage elements extracted from a vendor's library along with some combinational

```
use USER_UTILS.NEW_PROCS.PRESET_CLEAR;
-- libraries, other packages and entity as shown in Figure 5.5

-- edge-triggered synchronous process
architecture SYNC_WAIT of FOUR_SHIFT is
signal REG : HALF_REG;
begin
    process
    begin
        wait until (CLK'event and CLK='I');
        if (INIT='I') then
            REG <= X & REG(I to 3);
        else
            PRESET_CLEAR(REG, "I000");
        end if;
    end process;
    Z <= not REG(4);
end SYNC_WAIT;
```

Figure 5.8 Shift register architectures that use the Wait statement.

Box 5.7 How to create a synchronous section with a Wait statement

---

1 A process that contains a **Wait** statement is a template for a fully synchronous section of an architecture. The clarity of the code is often better than when **If** statements are used. The following syntax is required:

process
begin
  wait until clock_expression;
        -- sequential statements including signal assignments
  end process;

2 The valid clock_expressions are identical to those for an **If** construct but also include: (clock_signal=clock_value)

3 A sensitivity list cannot be used. The process is only sensitive to the signal in the clock expression.

4 The **Wait** must be the first statement in the process.

5 The sequential statements below the **Wait** *must* assign a value to a signal or to a previously undefined variable for the synthesizer to infer a flip flop and hence create a registered output. Otherwise, purely combinational logic will be generated.

6 Any combinational logic associated with the design must be placed in a separate process or be implemented using concurrent signal assignment statements.

7 Only one **Wait** statement can appear in each process.

---

logic. However, the designer will have had no control over the type of elements selected. The library or user-defined component can be instantiated in the usual way.

Figure 5.9 contains a structural shift register architecture constructed from D-type flip flops. Four component instantiation statements are used and the internal signals, Z and ZBAR are declared to facilitate interconnection. The external port signal X is fed into the D input of DFF1 while the output Y is derived from the inverted output of DFF4, QBAR. The parameter list determines the mapping of signals from the architecture to the components DFF1 and DFF2. Their mapping is achieved by positional association such that the first signal in the list is connected to the first port defined in the component's entity declaration, and so on. The component DFF3 makes the same associations by explicitly naming the source and target of each connection. The order in which these appear in the signal list is irrelevant. Component DFF4 mixes the two methods of creating a port map signal list. A parameter list is only one form of **interface list** that can use positional and named association. This and other lists are discussed in Box 5.8.

The behavioural description of the D-type flip flop, **DFF**, is given in Figure 5.10. This circuit employs an extra, internal signal, **INTERNAL_Q**, to store the state of D when a falling clock edge occurs. Outside of the process the external ports Q and QBAR are assigned their new states appropriately. This is not the shortest way of describing the behaviour of a simple D-type, which is left to the reader to determine (but see the Dos and Don'ts section in this chapter for possible pitfalls). The purpose of the signal **INTERNAL_Q** is to isolate the output ports of the architecture from the internal operations of the architecture. This is only really necessary in circuits where signal feedback is required, as will be seen shortly. It is also possible to use a variable inside the process that can be used to assign the new values to the outputs. It is, however, better to use a signal in this case as the Dos and Don'ts section explains.

Although it is feasible to create a small circuit with individual component instantiation statements, constructing port maps for large numbers of components is a tedious and error-prone exercise. If, as in this example,

```
entity FOUR_SHIFT is
    port  (
            X : in BIT;
            CLOCK, INIT : in BIT;
            Y : out BIT
            );
end FOUR_SHIFT;
```

```
-- manually structured D-type shift register
architecture STRUCTURAL of FOUR_SHIFT is
component DFF
    PORT (D, CLK, PC : in BIT; Q, QBAR : out BIT);
end component;
signal Z, ZBAR : BIT_VECTOR(I to 3);
begin
    -- indirect instantiations using positional and named association
    DFFI : DFF port map ( X, CLOCK, INIT, Z(I), ZBAR(I) );
    DFF2 : DFF port map ( Z(I), CLOCK, INIT, Z(2), ZBAR(2) );
    DFF3 : DFF port map ( D  =>Z(2),
                          CLK  =>CLOCK,
                          Q =>Z(3),
                          QBAR   =>ZBAR(3),
                          PC   =>INIT );
    DFF4 : DFF port map(Z(3), Q=>open, CLK=>CLOCK, PC=>INIT,
    QBAR=>Y );
end STRUCTURAL;
```

Figure 5.9 Shift register description using instantiated D-type flip flops.

Box 5.8 How to create an interface list

An **interface list** is the general name given to a list of interface constants, signals and/or variables that describe the information being exchanged between two sections of a design. The interface list will contain a series of **interface declarations** separated by commas. Interface lists can be found in:

- an entity declaration – a port list and/or generic list;
- a function/procedure declaration or port map – a parameter list;
- a component declaration – a port list and/or generic list;
- a block statement – a port list and/or generic list.

The general syntax for each type of declaration in a list is as follows:

```
interface_constant_declaration ::=
    [constant] identifier {,identifier}: [in] subtype_indication [:=expression]

interface_signal_declaration ::=
    [signal] identifier {,identifier}: [mode] subtype_indication [:=expression]

interface_variable_declaration ::=
    [variable] identifier {,identifier}: [mode] subtype_indication [:=expression]
```

File objects may also appear in an interface list but are not supported for synthesis.

If a mode is not specified, **in** is assumed. The possible modes are **in**, **out**, **inout** and **buffer**. The subtype indication must specify a synthesizeable type or subtype.

The optional expression supplies an initial value. This value will be ignored by the synthesizer unless a constant is being declared. Any expression must be globally static. It would be an error to give an initial value to any object that is not of mode **in** or **inout**.

Each type of list will allow some or all of the object declarations to be used.

- A port list can only contain signals of any mode. These define the interface between the entity or internal block and its external environment.
- A generic list can only contain constants. The constants provide information specific to a particular instance of an entity or internal block.
- A parameter list may contain all the objects. A procedure may send and receive values (modes **in**, **out** and **inout** supported) but a function only sends values in its list (mode **in** supported).

In all these lists it is possible to use named and positional association. Examples are given of each list throughout the examples in this book.

the structure is regular and array-like, the process of instantiating components can be left to a For-Generate statement. This uses an iteration scheme
to duplicate components and connect them in a cascading fashion. The
two forms that the For-Generate statement can take are illustrated in the
architectures REG_GEN and IRREG_GEN shown in Figure 5.11. Positional
association has been used exclusively in both these architectures.

The regular form of the construct, illustrated in REG_GEN, maps the same
interconnection structure into every discrete instantiation of a component.
The number of components that are instantiated is determined by the
number of iterations around the For-Generate loop. In this case, four JK-
type flip flops are cascaded together. In the regular structure, the connections to the ends of the array must be made by using concurrent signal
assignment statements outside the For-Generate statement. The syntax of
this form of the For-Generate statement is discussed in Box 5.9.

```
entity DFF is
   port (
           D : in BIT;
           CLK, PC : in BIT;
           Q, QBAR : out BIT
           );
end DFF;
```

-- D-type with synchronous clear, behavioural description
```
architecture BEHAVIOUR of DFF is
signal INTERNAL_Q : BIT;
begin
   -- outputs assigned concurrently (outside) the process
   Q <= INTERNAL_Q;
   QBAR <= not INTERNAL_Q;
   process
   begin
      wait until (CLK'event and CLK='0');
      if (PC='0') then
         INTERNAL_Q <= '0'
      else
         INTERNAL_Q <= D;
      end if;
   end process;
end BEHAVIOUR;
```

Figure 5.10 Behavioural description of a D-type flip flop with an additional
internal signal, INTERNAL_Q.

The architecture IRREG_GEN shows how an irregular array structure can still make use of the For-Generate statement. This example implements exactly the same shift register design as REG_GEN. A second level of conditional Generate statements is used to determine which port map should be selected for the nth component in the array. In this case, the input signals are mapped directly into the ports of the first component and the output signals taken directly from the output port of the fourth. The second and third flip flops are internal and use the same parameter list for port mapping as all the components used in the REG_GEN architecture. Note that expressions containing signals may not be passed to components. Hence, the extra internal signal K_IN is used to generate the inverse of X. This is then passed to the first flip flop.

Labels must be attached to each second-level For-Generate statement. See Box 5.10 for more details on the syntax of this form of the statement.

The component instantiated in both forms of the Generate statement is a JK-type flip flop. The behavioural description of this component is given in Figure 5.12. Like the D-type flip flop in Figure 5.10, an additional internal signal is used in the architecture. In this case, however, it is absolutely necessary to declare this signal in order that the present state of the JK flip flop may be stored. This state information is already available at the output port Q, but as it is not possible to *read* a port declared as an output, it must be stored internally using another signal. S is the signal declared inside the architecture for this purpose.

The fully synchronous process triggers on a falling-edge clock signal. If the reset signal PC is not active, the new state of S is then determined within the Case statement. The new state of S is used to update the outputs Q and QBAR in concurrent signal assignment statements that precede the process. Remember that the order of concurrent statements is irrelevant to their order of execution.

5.2.3 Synthesis and optimization

Preliminaries

All the descriptions of the 4-bit shift register are compiled into a generic component library. This contains a range of standard flip flop elements with features such as asynchronous preset and clear inputs. When inferring sequential elements the synthesizer will usually select the simplest storage element – usually a D-type. Hence, most of the synthesis results presented here contain one or more of these elements. Of course, when instantiating components, the choice of flip flop – D, JK or SR type – is totally a decision for the designer. If a particular technology has already been selected then the choice of component will come from the vendor's libraries. In the structural style examples here, the components, DFF and

```
-- entity as shown in Figure 5.9

--regular For-Generate statement
architecture REG_GEN of FOUR_SHIFT is
component JKFF
  PORT (J, K, CLK, PC : in BIT; Q, QBAR : out BIT);
end component;
signal Z, ZBAR : BIT_VECTOR(0 to 4);
begin
  SHIFTER : for N in 1 to 4 generate
    -- any concurrent statement can be located inside a generate
statement
    JKFFN : JKFF port map (Z(N-1), ZBAR(N-1), CLOCK,
                                   INIT, Z(N), ZBAR(N) );
  end generate SHIFTER;
  Z(0) <= X;
  ZBAR(0) <= NOT X;
  Y <= ZBAR(4);
end REG_GEN;

-- irregular For-Generate statement
architecture IRREG_GEN of FOUR_SHIFT is
component JKFF
  PORT (J, K, CLK, PC : in BIT; Q, QBAR : out BIT);
end component;
signal Z, ZBAR : BIT_VECTOR(1 to 3);
signal K_IN : BIT;
begin
  -- labels for each For-Generate statement are compulsory
  SHIFTER : for N in 1 to 4 generate
    FIRST : if (N=1) generate
        JKFFN : JKFF port map ( X, K_IN, CLK, INIT, Z(1), ZBAR(1) );
    end generate FIRST;
    MIDDLE : if (N>1 and N<4) generate
        JKFFN : JKFF port map (    Z(N-1), ZBAR(N-1), CLK, INIT,
                                   Z(N), ZBAR(N) );
    end generate MIDDLE;
    LAST : if (N=4) generate
        -- 'open' keyword indicates no connection to that port
        JKFFN : JKFF port map ( Z(3), ZBAR(3), CLK, INIT, open, Y );
    end generate LAST;
  end generate SHIFTER;
  K_IN <= NOT X;
end IRREG_GEN;
```

Figure 5.11 Regular and irregular Generate statement architectures for creating the 4-bit shift register.

Box 5.9 How to use a regular For-Generate statement

The Generate statement allows a section of a concurrent VHDL description to be replicated using an iterative scheme. It is frequently applied to the creation of multiple instances of the same component in a structural style architecture or used to create circuits with any regular structure. The iteration parameter is used as an index to create unique instances of each statement.

When replicating the same set of concurrent statements in an exact form, a regular For-Generate statement can be used by applying the syntax:

```
label : for iteration_parameter in value1 direction value2 generate
[ generate_declarative_part
begin ]
   concurrent_statements     --such as component instantiations
end generate [label];
```

1 The label at the start of the generate statement is mandatory.
2 The iteration parameter (in any For statement) can be considered to be a constant object. It is implicitly declared.
3 Value1 and value2 must be static, discrete values (probably integers). The direction is To for an ascending range and Downto when descending.
4 The generate declarative part may contain local declarations (such as an architecture declarative part) used by the concurrent statements.
5 Multidimensional structures can be created by nesting For-Generate statements. Structures with more than two dimensions are not recommended.

JKFF are user-defined versions of flip flops that are compiled into the working library. The compile-time options are effectively the same as those shown in section 4.2.3 (barring specific names).

Statistics for each circuit are again presented in this section. However, in a sequential circuit it is not possible to determine the input to output delay as this is dependent upon the frequency of the flip flops' clocking signal. Instead it is more appropriate to look at one of the following:

• input to latch delay
• latch to output delay
• latch to latch delay.

Box 5.10 How to use an irregular For-Generate statement

An irregular Generate statement uses If-Generate statements nested
in the For-Generate to create a more flexible iteration scheme. This
allows, for example, the first and last (or indeed any!) components in
an array to have a unique interconnection scheme. The syntax for an
irregular Generate statement is as follows:

```
label : for iteration_parameter in value1 direction value2 generate
[ generate_declarative_part
begin ]
   {labelN : if conditionN generate
   [ generate_declarative_part
   begin ]
      concurrent_statements
   end generate [ labelN ];}
end generate [ label ];
```

1 The condition attached to each If-Generate statement must be static.
   It will therefore usually test the integer value of one or more iteration parameters. For example:

$$X > 3 \text{ and } Y < 2$$

2 For-Generate statements may be nested inside If-Generates and vice
   versa. If-Generates may be nested inside other If-Generates.
3 See Box 5.9 for additional comments.

The choice is dependent upon the number of latches in the circuit and
their position in relation to any additional combinational logic. When circuits are being directly compared in this text the same statistic is presented.

Logic synthesis and optimization results

Synthesis of the behavioural architectures
Each of the behavioural architectures consists of a process that describes a
particular variant of the 4-bit shift register and an output inverter. These
are two concurrent statements that will appear at the top level of the synthesized circuit. All behavioural architectures except SIMPLE employ an
external initialization signal called INIT and therefore only SIMPLE has a
different top-level circuit. Figure 5.13a shows the top level of SIMPLE and
Figure 5.13b shows the same for all the other behavioural architectures. In
each case the block represents the Process statement, the output of which is
fed into an inverter.

```
entity JKFF is
   port  (
      J, K : in BIT;
      CLK, PC : in BIT;
      Q, QBAR : out BIT
      );
end JKFF;

architecture BEHAVIOUR of JKFF is
signal S : BIT;
begin
   Q <= S;
   QBAR <= not S;
   process
   variable JK : BIT_VECTOR(0 to 1);
   begin
      wait until (CLK'event and CLK='0');
      JK := J & K;
      if (PC='0') then
         S <= '0';
      else
         case JK is
            when "00" => S <= S;
            when "10" => S <= '1';
            when "01" => S <= '0';
            when "11" => S <= not S;
         end case;
      end if;
   end process;
end BEHAVIOUR;
```

Figure 5.12 JK flip flop behavioural description.

Below the top level (inside the process), most of the synthesized circuits are different. The process in **SIMPLE** constructs the circuit shown in Figure 5.14. This is the most basic demonstration of how a signal assignment inside a process with a clock expression template will infer flip flops. The synthesizer has used the most basic storage elements available – edge-triggered D-types.

The processes inside **STRING_ASSIGN** and **AGGREGATE** allow synchronous initialization of REG. The former uses a bit-string literal in its assignment and the latter an aggregate. This has made no difference to the circuits produced by the synthesizer for each, which are identical and shown in Figure 5.15. Note that using a std_ulogic type for the signal CLK in AG-

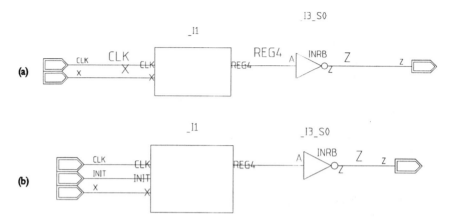

Figure 5.13 Top-level synthesized circuits for the architectures (a) SIMPLE, (b) STRING_ASSIGN, AGGREGATE, ASYNC_IF1, ASYNC_IF2 and SYNC_WAIT.

GREGATE has also made no difference. This is related to how the signal has been used in this circuit. Generally, designing with multivalued logic will affect how the code has to be written. This is discussed further in Chapter 7.

As the figure shows, a synchronous initialization process means that when INIT is high, the value of REG is shifted through as usual, but when low the value shifted onto the outputs of the flip flops on the next clock cycle will be 1000.

The SYNC_WAIT architecture demonstrated how a Wait statement can be used to infer a sequential section. The whole process must be synchronous and therefore only synchronous initialization is possible with a Wait statement. The circuit produced by this form of sequential logic template is

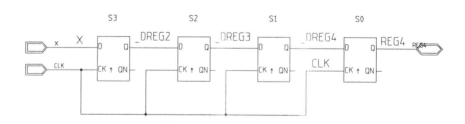

Figure 5.14 Synthesized circuit described by the process in the SIMPLE architecture.

identical to the one produced by the If statement in the **STRING_ASSIGN** and **AGGREGATE** architectures. This circuit is therefore also shown in Figure 5.15

Table 5.1 shows a comparison of the synthesis statistics for these architectures with the **SIMPLE** architecture, which contains no initialization function. In terms of transistors, synchronous initialization makes the basic circuit approximately one-third larger. The additional gate between each flip flop also increases the latch to latch delay.

The **ASYNC_IFI** and **ASYNC_IF2** architectures demonstrated two methods for asynchronous initialization of the flip flops. Their synthesized circuits are shown in Figures 5.16 and Figure 5.17 respectively. The former seems to be a very complex implementation of what should be a simple circuit. In fact it is a very clear demonstration of the effects of an incomplete If statement. Although both conditions of the INIT signal are covered in the process, they are not covered in the same If statement. This is not a problem in the combinational section when REG is being initialized. However, in the synchronous section, what state is the circuit supposed to assume if INIT is low when the circuit is clocked? As it is not told, the synthesizer uses the present state by feeding back the Q of each flip flop. The SD input selects the data source, D0 or D1, depending on the value of INIT. However, the value on D0 is never actually used as asynchronous initialization occurs when INIT is low. Incomplete If statements are discussed further in section 5.4.3 of the Dos and Don'ts section.

The better approach to asynchronous initialization is demonstrated in the **ASYNC_IF2** architecture and the synthesized process shown in Figure 5.17. The If statement is completed in this case by the clocking condition and therefore a much simpler circuit is produced. The only function of the INIT signal in this case is to feed the appropriate asynchronous input of each flip flop. The synthesis statistics for these two circuits in Table 5.2 indicate that the latter is roughly one-third smaller than the former.

As would therefore be expected, the statistics in Table 5.2 show that **ASYNC_IF2** has latch to latch delays that are approximately one-third lower

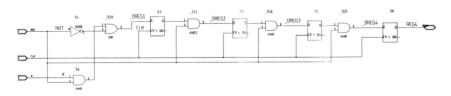

Figure 5.15 Synthesized circuit described by the process in both the STRING_ASSIGN, AGGREGATE and SYNC_WAIT architectures.

Table 5.1 Logic synthesis and optimization statistics for four synchronous dataflow-style architectures with and without external initialization signals. Compares If and Wait versions

Parameters Temperature: 25.0 °C Voltage: 5.0 V	Synthesis Library: generic	Optimization Type: area Level: low
Circuit: FOUR_SHIFT-SIMPLE		
Cells	5	4
Transistors	74	72
Area	9506	9154
Longest latch to latch delay	1.659	No change
Circuit: FOUR_SHIFT-STRING_ASSIGN FOUR_SHIFT-AGGREGATE FOUR_SHIFT-SYNC_WAIT		
Cells	11	9
Transistors	106	98
Area	13379	12322
Longest latch to latch delay	2.297	No change

and requires a third fewer transistors than ASYNC_IF1. The additional complexity of the flip flops in the first version of the circuit are the cause of this, demonstrating the cost of poor design methods and, in particular, the incomplete If statement.

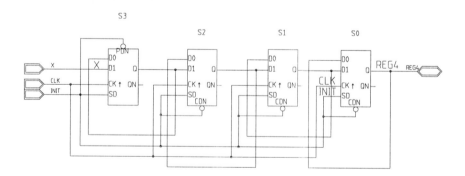

Figure 5.16 Synthesized circuit described by the process in the ASYNC_IF1 architecture.

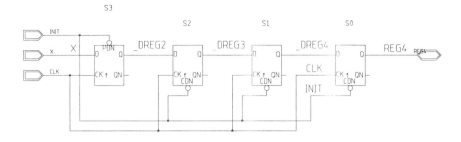

Figure 5.17 Synthesized circuit described by the process in the **ASYNC_IF2** architecture.

Optimization of the behavioural architectures

Optimization only involved the lowest level of effort in an area reduction process. This aimed simply to strip out redundant and duplicated logic in the shortest time.

One of the most obvious redundant logic elements in all the circuits seen so far has been the inverter on the output. In every case, this output can be obtained directly from the inverted output, QN, of the final flip flop. To remove this component through optimization the process block must be flattened. The resulting circuit for the SIMPLE architecture is not shown but, as Table 5.1 indicates, a cell containing two transistors has been optimized out of the circuit.

The optimized circuits for **STRING_ASSIGN**, **AGGREGATE** and **SYNC_WAIT** are (thankfully!) identical and shown in Figure 5.18. This shows

Table 5.2 Logic synthesis and optimization statistics for the two asynchronous behavioural style architectures

Parameters	Synthesis	Optimization
Temperature: 25.0 °C	*Library: generic*	*Type: area*
Voltage: 5.0 V		*Level: low*
Circuit: FOUR_SHIFT-ASYNC_IFI		
Cells	5	10
Transistors	138	124
Area	15843	15491
Longest latch to latch delay	2.775	3.082
Circuit: FOUR_SHIFT-ASYNC_IF2		
Cells	5	No change
Transistors	92	90
Area	11794	11618
Longest latch to latch delay	1.813	No change

the output Z being derived from QN. The optimizer has also performed some straightforward Boolean reduction on the logic at the input to the circuit. This is shown in equation 5.1 (note that the signal _DREG1 is re-named _I1__DREG1 by the optimizer).

$$
\begin{aligned}
\_DREG1 &= \overline{INIT} + X.INIT \\
&= \overline{INIT}.X + \overline{INIT}.\overline{X} + INIT.X \\
&= \overline{INIT}.\overline{X} + \overline{INIT} + X
\end{aligned}
\tag{5.1}
$$

The optimization statistics in Table 5.1 for all three architectures also con-firm that they are identical.

The optimized circuits for the asynchronous architectures are, however, quite different. ASYNC_IF1, shown in Figure 5.19, has undergone a dra-matic transformation that has doubled its cell count but actually reduced the number of transistors. This has been achieved by replacing each 'all-in-one' multiplexer and flip flop element with simpler D-type flip flops and front-end combinational logic. The flip flops still retain their asynchronous initialization inputs and feedback connection but the multiplexed data in-puts have gone. Note that the feedback connection now uses the inverted output of each flip flop. The result of area optimization is a slightly smaller

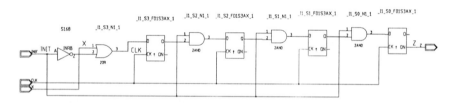

Figure 5.18 Optimized circuit for the STRING_ASSIGN, AGGREGATE and SYNC_WAIT architectures. Each two-level hierarchy has been flattened.

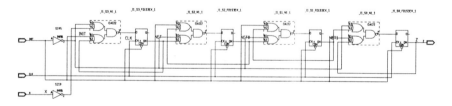

Figure 5.19 Optimized circuit (with flattened hierarchy) for the ASYNC_IF1 architecture.

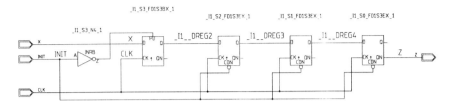

Figure 5.20 Optimized circuit (with flattened hierarchy) for the ASYNC_IF2 architecture.

circuit but, as quite often happens, it is slower. In fact, this circuit has the longest latch to latch delay of all the behavioural style architectures.

When ASYNC_IF2 is optimized its circuit is much smaller than ASYNC_IF1 but is relatively unchanged from its original version. However, the changes that have been made are interesting. As Figure 5.20 shows, the output inverter has of course been stripped out but INIT has now been inverted. This connects to the preset input of the first flip flop, which has been swapped for one that is active high. As the statistics in Table 5.2 indicate, the combination of external inverter with active high preset requires two fewer transistors than simply an active low preset. This result is very much related to the library in use but it demonstrates the power of the optimizer for determining the best combination of components. Of course, such optimization relies on the target library containing a good selection of different flip flops that make this choice possible in the first place.

Synthesis of the structural architectures' components
This section looks at the components used in the structural style architectures of the shift register. These components are constructed in a behavioural style and describe D and JK-type flip flops. Both architectures, DFF and JKFF respectively, contain a synchronous reset, PC, and employ a Wait statement.

The top-level circuit for the DFF architecture is shown in Figure 5.21a and the implementation of the architecture's process (the contents of the block) in Figure 5.21b. As stated in the architectural description, the output from the block is INTERNAL_Q and Q and QBAR are then derived from that – hence the existence of the top-level inverter. Inside the process block the circuit is as should be expected apart from the redundant, duplicated inversion logic. This is probably an anomaly of the synthesizer that will be cleared up by the optimizer.

Figure 5.22 shows the same two levels of the JKFF architecture. Again this double inversion operation is present on the PC signal but this is not the most concerning aspect of the process' implementation. This circuit

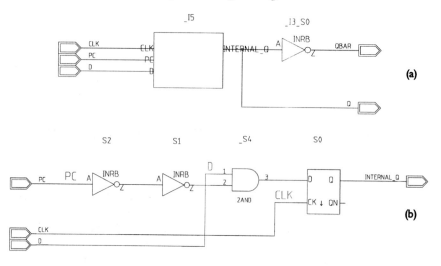

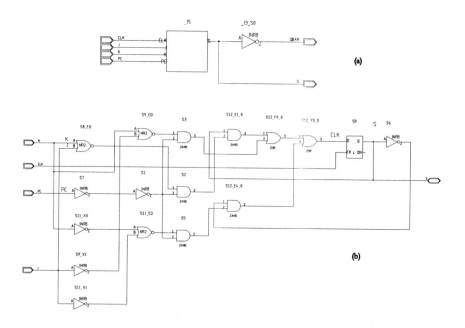

Figure 5.2.1 Synthesized **DFF** architectural description. (a) Top-level circuit. (b) Circuit described by the **Process** statement.

Figure 5.22 Synthesized **JKFF** architectural description. (a) Top-level circuit. (b) Circuit described by the **Process** statement.

produced does not contain a single JK flip flop. Instead, a section of combinational logic has been glued to the front end of a D-type. Why?

This occurs because the synthesizer does not 'see' a JK flip flop in the process' function but knows, from the clock condition in the Wait statement, that a sequential circuit is being inferred. As discussed in section 5.1.3, the choice of flip flop in an inferred sequential section is left entirely up to the synthesizer. It will therefore tend to select a simple one rather than a complex one, and so not risk the overspecification of the circuit. Observe the feedback of S, which is required to determine the next state of the circuit. Because this signal is not only updated and output by the process but also read, the connection to the top level of the circuit is defined in the schematic as bidirectional. This does not mean that it can be used as an input to the circuit as the entity declaration for this component states that the port Q, to which it is directly connected, is mode output.

The statistics for these two circuits are shown in Table 5.3. As expected, they indicate that the circuit for the JK flip flop is three times as large as the one for the D-type. The delay figures given in this table compare the effects of the front-end combinational logic on each circuit. They indicate that the longest delay for an input signal to reach the latch in the JK flip flop is well over double that of the D-type.

Optimization of the structural architectures' components
The flattened, optimized version of the D-type flip flop has had its redundant logic stripped out as was hoped. This includes the removal of the double inversion and the use of the QN output on the flip flop. Figure 5.23 illustrates the circuit.

The optimized JK flip flop has had this same logic removed, but also much more. In fact, as Figure 5.24 illustrates, it is almost beginning to re-

Table 5.3 Logic synthesis and optimization statistics for the D- and JK-type user-defined components

Parameters Temperature: 25.0 °C Voltage: 5.0 V	Synthesis Library: generic	Optimization Type: area Level: low
Circuit: DFF-BEHAVIOUR		
Cells	5	2
Transistors	30	24
Area	4048	2992
Longest input to latch delay	1.942	1.304
Circuit: JKFF-BEHAVIOUR		
Cells	18	4
Transistors	86	34
Area	11266	4224
Longest input to latch delay	4.214	2.294

semble a JK-type flip flop (despite the presence of a D-type!). The Boolean equation for the front-end logic is given in equation 5.2. The And gate has been excluded from this equation as it simply allows the reset signal PC to control the input to the flip flop. If the equation is manipulated a little, the well-known equation of a JK flip flop emerges, as shown in equation 5.3. Of course, it would be preferable to infer a JK flip flop directly in the first place but the reasons for not doing this have been covered earlier in the chapter.

$$Z = \overline{(K + \overline{Q}).(\overline{J} = Q)} \qquad\qquad (5.2)$$

$$
\begin{aligned}
Z &= \overline{(K + \overline{Q})} + \overline{(\overline{J} + Q)} \\
&= \overline{\overline{K}}\,\overline{Q} + \overline{\overline{J}}\,\overline{\overline{Q}} \\
&= J\overline{Q} + \overline{K}Q
\end{aligned}
\qquad\qquad (5.3)
$$

The optimization statistics in Table 5.3 show that the JK flip flop has had almost two-thirds of its logic stripped out whereas the transistor count for the D-type has only fallen by one-fifth. This demonstrates the highly circuit-dependent nature of the optimizer's success and that an efficient VHDL description does not guarantee that the synthesizer will produce a 'good' circuit.

Synthesis of the structural architectures

The structural architectures demonstrated the use of manual and generated component instantiations in the creation of the 4-bit shift register. A component instantiated from a vendor's library will always be optimized. It is therefore appropriate in these examples to instantiate the optimized versions of the D- and JK-type flip flops.

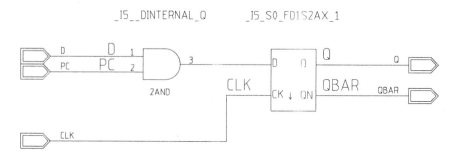

Figure 5.23 Optimized DFF architecture with a flattened hierarchy.

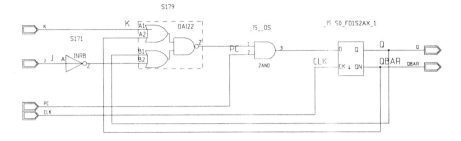

Figure 5.24 Optimized JKFF architecture with a flattened hierarchy.

Figure 5.25a shows the top-level circuits for the **STRUCTURAL** architecture containing the D-type components and Figure 5.25b the **REG_GEN** and **IRREG_GEN** architectures containing JK-types. The main difference between the two circuits is the need for the second input **K** to be provided in Figure 5.25b, hence the extra inverter at the front of the circuit. This inverter is also the only point in the circuit where REG_GEN differs from IRREG_GEN. As the circuit figure shows, there are two names for the signal output from the inverter, **ZBAR0** in REG_GEN and **K_IN** in IRREG_GEN (shown in parentheses). The reason for this is obvious from the architectural descriptions of the two circuits.

As the statistics show in Table 5.4, the area parameters of the synthesized circuit are simply four times optimized components' parameters. The delays given for these circuits are latch to latch and not input to latch as for the components; hence the difference.

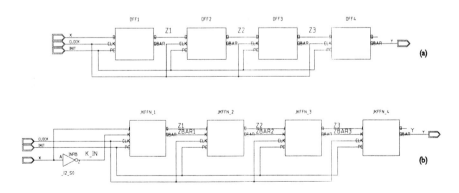

Figure 5.25 Synthesized (a) STRUCTURAL (b) REG_GEN and IRREG_GEN architectures showing the instantiated flip flop components.

Table 5.4 Logic synthesis and optimization statistics for the structural style architectures. Manually instantiated D-type and regular/irregular generated JK-type structures. All components are the optimized versions

Parameters Temperature: 25.0 °C Voltage: 5.0 V	Synthesis Library: generic	Optimization Type: area Level: low
Circuit: FOUR_SHIFT-STRUCTURAL		
Cells	8	No change
Transistors	96	No change
Area	11970	No change
Longest latch to latch delay	2.684	No change
Circuit: FOUR_SHIFT-REG_GEN FOUR_SHIFT-IRREG_GEN		
Cells	17	13
Transistors	138	122
Area	17251	15139
Longest latch to latch delay	4.353	4.240

Optimization of the structural architectures

The use of optimized components in these circuits does not necessarily mean that further optimization cannot take place. While each component may be optimized, a global picture of the circuit, which contains several identical components, may indicate redundancy across components. In a real implementation that uses a particular technology library, components will often represent hardware macros. In this case it may not be possible to flatten components and remove these redundancies. Here, the effect of being able to remove such redundancy is demonstrated.

Optimization of the STRUCTURAL architecture has not produced any improvement in the circuit. The simple nature of the components and in particular the isolation of combinational logic between the sequential logic has inhibited any further minimization. In the REG_GEN and IRREG_GEN architectures, however, optimization has made area savings of over 10% (Table 5.4). These occur because, as the number of identical components increase, it becomes more worthwhile to make changes that allow common elements to be extracted. The optimized circuit is shown in Figure 5.26. As this shows, only the final stage (combinational logic and flip flop) of the circuit remains unchanged. This occurs because the optimizer has worked its way through the circuit looking for redundancy, etc. When it reaches the final stage it can look no further and so the component remains in its previously optimized state.

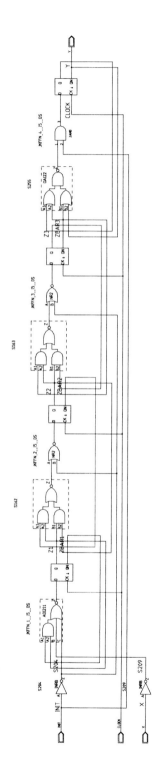

Figure 5.26 Optimized REG_GEN and IRREG_GEN architectures with flattened hierarchy.

5.2.4 Shift register summary

The shift register example has demonstrated how If and Wait statement templates can be used to infer sequential logic in a VHDL description. The format of these templates is reasonably fixed and signals and variables within the sequential logic must adhere to certain guidelines. In particular, the effect of creating incomplete If statements was demonstrated as one serious pitfall that should be avoided.

The structural architectures demonstrated that there are efficient ways of replicating sections of concurrent statements using the For-Generate statement. Often this is used to instantiate components but can be used for any concurrent statement. Even though two of these structural architectures instantiated JK flip flop components, these were user defined and the behavioural description of the JK's operation was implemented by the synthesizer using combinational logic and a D-type. This highlighted the point that the designer never really knows what the synthesizer will produce from a behavioural architecture. If more control is required, flip flops of the appropriate type must be instantiated directly from a (vendor's) library.

5.3 UP–DOWN PRESETABLE COUNTER

This section presents a more practical design example, employing some of the language constructs and features illustrated in this chapter. A behavioural design style has been adopted for this circuit, an up–down counter with a parallel loading facility. The counter is only 3 bits wide so that its characteristics can be observed without the needless repetition of logic.

5.3.1 Functional overview

The counter has the following features (and external I/O):

- 3-bit fully synchronous operation;
- rising edge triggered (by CK);
- count up (CUCD='0') or count down (CUCD='1');
- continuous loop operation;
- synchronous parallel load facility (LOADVAL read in when LOAD='0');
- counter enabling signal (ENABLE), which suspends the counting operation but not the parallel loading;
- carry out signal (CO), which goes high when either counting up and the counter loops round to 0 or when counting down and it loops round to its maximum value.

5.3.2 VHDL code design

The aim with this VHDL description was to make it as flexible as possible. With this in mind, the architecture was designed so that with the change of a single parameter the size of the counter could be increased or decreased. Another requirement was to separate the sequential and combinational sections in the description to make it more understandable. The resulting VHDL code has two processes and uses a number of constants. The design unit for the entity declaration is shown in Figure 5.27. This also contains a Library and Use clauses to access packages. The DZX library is vendor specific and provides functions that enable the design of the counter to be simplified. The functions that are used are:

- an adder function +. This performs the arithmetic addition of two BIT_VECTORs and returns a BIT_VECTOR;
- a subtraction function –, this performs the arithmetic subtraction of two BIT_VECTORs and returns a BIT_VECTOR;
- a clock detection function RISING_EDGE, this take a signal of type BIT and returns TRUE when a rising edge is detected; this function differs from the one used in the std_logic_1164 package, which takes a signal of type std_ulogic.

Every synthesis tool vendor will undoubtedly supply the same or very similar functions and so their use here is justified. It also enables the subject of operator and function overloading to be introduced by an example before it is discussed in detail in the next chapter.

The identifiers given in the Use clauses are all item specific. Hence, only the two required operator functions in the BIT_ARITH package and the clock function in the BIT_UTILS package are visible.

The architectural description of the counter is shown in Figure 5.28. To change the size of the counter, the only change required is to the constant CBITS in the declarative part. Note that an appropriate change would also

```
library DZX;
use DZX.BIT_ARITH."+", DZX.BIT_ARITH."-";
use DZX.BIT_UTILS.RISING_EDGE;
entity UP_DOWN is
    port (
      LOADVAL : in BIT_VECTOR(2 downto 0);
      CK, CUCD, ENABLE, LOAD : in BIT;
      CO : out BIT
      );
end UP_DOWN;
```

Figure 5.27 Libraries, packages and entity of the up–down presetable counter.

```
architecture BEHAVIOUR of UP_DOWN is
  -- set up the size of the counter
  constant CBITS : POSITIVE := 3;
  constant MAXVAL : BIT_VECTOR(CBITS-I downto 0) := (others=>'I');
  constant MINVAL : BIT_VECTOR(CBITS-I downto 0) := (others=>'0');
  constant STEPVAL : BIT_VECTOR(CBITS-I downto 0) := (0=>'I',others=>'0');
  signal COUNTVAL : BIT_VECTOR(CBITS-I downto 0);
begin

  COUNTING : process
    variable TEMPCOUNT : BIT_VECTOR(CBITS-I downto 0);
  begin
    wait until RISING_EDGE(CK);
    TEMPCOUNT := COUNTVAL;        -- initialize the variable
    if LOAD='0' then          -- parallel load the counter
      TEMPCOUNT := LOADVAL;
    elsif ENABLE='0' then            -- counter is enabled
      if CUCD='0' then            -- count up
        if TEMPCOUNT=MAXVAL then
          TEMPCOUNT := MINVAL;
        else
          TEMPCOUNT := "+" (TEMPCOUNT, STEPVAL);
        end if;
      else                -- count down
        if TEMPCOUNT=MINVAL then
          TEMPCOUNT := MAXVAL;
        else
          TEMPCOUNT := TEMPCOUNT - STEPVAL;
        end if;
      end if;
    end if;
    COUNTVAL <= TEMPCOUNT;      -- put the counter value back
  end process COUNTING;

  CARRY : process (CUCD, COUNTVAL)
  begin
    if  (CUCD='0' and COUNTVAL=MINVAL) or
        (CUCD='I' and COUNTVAL=MAXVAL) then
      CO <='I';              -- send out carry bit
    else
      CO <='0';
    end if;
  end process CARRY;
end BEHAVIOUR;
```

Figure 5.28 Architectural description of the up–down presetable counter.

be required to the number of bits in LOADVAL in the Port statement of the entity. Another way to create a size-independent architecture would be to use a Generic statement in the entity declaration. This would contain a constant, such as CBITS, which would also determine the size of LOADVAL. CBITS would be visible to the declarations in the architecture body. This approach is not used here as generics are not always supported for synthesis.

The value of CBITS could not be supplied by an external signal as the size of the counter must be known at compile time. Hence, CBITS must be static and therefore a constant.

The other constants set up the maximum and minimum value of the counter and the step value, all using CBITS to determine their size. The values of each are assigned through aggregates. The signal COUNTVAL is declared and will be used as the register that stores the present counter value. No initial value is assigned to this signal.

The main process is fully synchronous and labelled **COUNTING**. Its function can be followed through quite easily but its key feature is in the use of a variable to store the counter value. When a rising edge is detected, the sequence of operations in this process is as follows.

1 Assign the counter value to the variable. This ensures that the variable is always initialized before it is used and therefore prevents unwanted flip flops.
2 Update the variable in a manner appropriate to the present state of the control signals.
3 Assign the new value of the counter stored in the variable back to the signal. The signal therefore infers sequential logic as it is assigned inside a synchronous section.

This technique of passing signals to variables to be modified in a process is valuable because the variable updates occur immediately. This makes the flow of execution much clearer and avoids hazards. As signal updates all occur together at the end of the process and there is always only one update per signal.

The assignment statements in the process that increment and decrement the counter value illustrate the two ways that an overloaded operator can be used. The addition operator is use like a function call and the subtraction operator is used like a standard operator. Operator overloading is discussed further in Chapter 6.

The second process labelled **CARRY** determines whether the carry out signal **CO** should be set or reset. This is a purely combinational logic description, which could have been implemented using a dataflow style. Instead a process was used to keep the design consistent and modular.

5.3.3 Synthesis and optimization

Depending on the number of bits selected, this counter has the potential to be very large and repetitive in its implementation. Even with only 3 bits, displaying the basic synthesized circuit in its entirety would be too complex to follow and uninformative for the reader. Only selective elements are therefore illustrated.

Preliminaries

The logic library name **DZX** should be linked to the physical **DZX** library to enable the two packages **BIT_ARITH** and **BIT_UTILS** to be accessed. Some of the compile-time options for the circuit include the following:

Directory and filename: ./udcounter.vhdl
Type of input: VHDL
Entity name: UP_DOWN
Architecture name: BEHAVIOUR
Share resources: yes
Destination library: generic

The share resources option has been set to 'yes' in this example. The resources in the circuit include the circuit that will be created to implement the adder and the circuit for the subtractor. This is quite an overhead if two similar circuit functions are created with separate logic. By allowing resource sharing the logic requirements can be reduced. It is usually preferable to allow sharing. This subject will be discussed further in the context of arithmetic circuits in the next chapter.

Synthesis results

The synthesized circuit consists of three circuits on two levels of the design hierarchy. The architecture itself creates the top level. Inside this there are two concurrent statements – the processes. Each of these creates its own block as Figure 5.29 shows. The inputs to the first block are all the signals read by the COUNTING process whereas the updated signals in this block are output and then read by the second block. This represents the CARRY process. Any signal in this process' sensitivity list will be an input to the block (assuming it is read).

The synthesized contents of these processes are not shown but their statistics are given in Table 5.5. As the timing statistics indicate, the COUNT-ING process contains clocked storage elements whereas CARRY contains only combinational logic. The maximum frequency of the counter has been shown here to enable a clear comparison to be made between this and the optimized circuit.

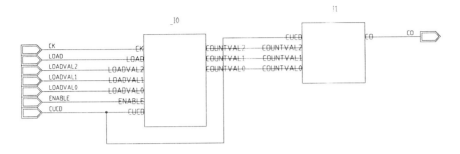

Figure 5.29 Synthesized top-level circuit of the up–down counter's architecture, BEHAVIOUR

Optimization results

The BCD to seven-segment display converter designed in the previous chapter was optimized in two ways. One involved the global optimization of the circuit and the flattening of its component hierarchy. The other preserved the hierarchy and optimized individual components. There was shown to be little to choose between either approach.

As this example has two clearly distinct sections, one combinational and one sequential, the hierarchy has been preserved and each process optimized independently. Again only area optimization has been attempted to remove simply the redundant or duplicated logic. Low, medium and high levels of effort were employed.

The optimization statistics for the two processes are shown in Table 5.6. As with other examples, the majority of the unnecessary logic is stripped out with the lowest level of effort. In fact, the size of the CARRY process' logic is such that further optimization achieves no improvement in circuit characteristics.

Table 5.5 Logic synthesis statistics for the process blocks COUNTING and CARRY in the architecture BEHAVIOUR of the up–down counter.

Parameters	Synthesis	
Temperature: 25.0 °C	Library: generic	
Voltage: 5.0 V		
Circuit: UP_DOWN-BEHAVIOUR	COUNTING process	CARRY process
Cells	45	9
Transistors	264	38
Area	33447	49290
Longest latch to latch delay /		
Maximum frequency	7.778 ns / 128 MHz	Not applicable
Longest input to output delay	Not applicable	2.937 ns

Table 5.6 Logic optimization statistics for the process blocks COUNTING and CARRY in the architecture BEHAVIOUR of the up–down counter

Parameters	Optimization		
Temperature: 25.0 °C	Type: area		
Voltage: 5.0 V	Level: low	medium	high
Circuit: UP_DOWN-BEHAVIOUR			
COUNTING process			
Cells	26	No change	21
Transistors	176	No change	166
Area	22357	No change	20068
Longest latch to latch delay/	8.437 ns/	No change	7.529 ns/
maximum frequency	118 MHz		132 MHz
CARRY process			
Cells	5	No change	No change
Transistors	24	No change	No change
Area	2992	No change	No change
Longest input to latch delay	1.793 ns	No change	No change

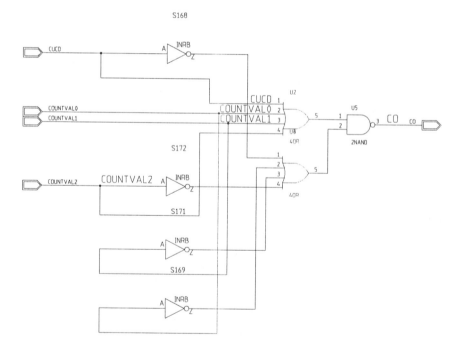

Figure 5.30 Optimized circuit for the CARRY process of the up–down counter.

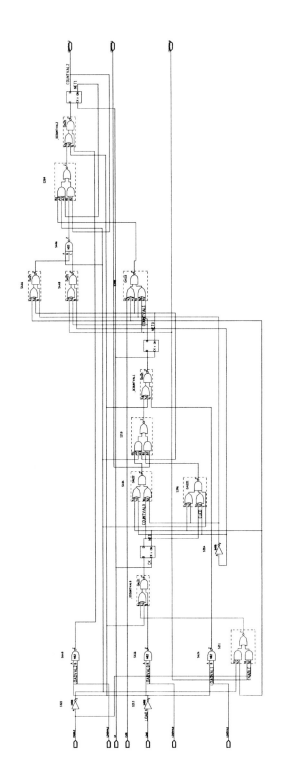

Figure 5.31 Optimized circuit for the COUNTING process of the up–down counter.

The initial optimization of the COUNTING process causes a significant reduction in transistor count and therefore area but, as often occurs, the latch to latch delay increases. The high level of optimization effort has produced a final reduction in area but the delay has only been reduced slightly below its original value. Although not attempted here, timing optimization should now be performed on this circuit and some of the area gains traded off for improvements in the speed.

The schematic diagrams of the fully optimized CARRY and COUNTING process block circuits are shown in Figures 5.30 and Figure 5.31 respectively. In the latter the three rising edge-triggered flip flops that store the present state of the counter, COUNTVAL, can be clearly seen. Each bit of the counter is output through bidirectional connectors (as they are read as well as updated). Also all bits, except the most significant, are fed to the combinational logic that generates the flip flop input for the next bit.

5.3.4 Counter summary

The counter constructed here can be modified easily for larger or smaller implementations. Using constants in the ways shown allows a flexible and clear description to be generated. The example has also demonstrated how the external control features of such a counter are implemented. Although this was a fully synchronous counter, features such as the parallel loading could be made asynchronous with only simple modifications. The technique of using variables inside a process to represent signals is a useful one. As the control logic examples in Chapter 7 will demonstrate, it is a valuable approach to adopt for finite-state machine design.

5.4 DOS AND DON'TS

5.4.1 Selecting subtype or type

A subtype is a type with a constrained range. The type referred to here is known as the subtype's **base type**. However, a base type may itself be a subtype:

subtype SHORT **is** INTEGER **range 0 to** 127;
subtype MAGNITUDE **is** SHORT **range 0 to** 63;

In these two examples the base types are INTEGER and SHORT respectively.

Having defined a subtype, it is possible to use it as a substitute in any subprogram parameter list or in an assignment statement, for example, that uses its base type or even the base type of the base type, and so on. Effectively the subtype is the same type as its base type. Of course, it is important to ensure that when assigning values to objects of a particular subtype the source (either a literal or an object of the same type) is within the range defined for the subtype.

A type definition, on the other hand, creates a new base type and may be constrained or unconstrained. Consider the subtype defined in the package NEW_PROCS shown in Figure 5.2. This is a subtype of the unconstrained base type BIT_VECTOR. It may therefore use any function defined for that type or be used in assignment statements with other BIT_VECTOR subtypes. This was done in the SYNC_WAIT architecture when the 4-bit BIT_VECTOR REG was passed to the PRESET_CLEAR procedure. The following declaration also creates a 4-bit array of type BIT:

type HALF_REG **is array(**I **to** 4**) of** BIT;

However, this is not a BIT_VECTOR subtype, it is the new type HALF_REG, which just happens to be composed of the same element type as a BIT_VECTOR.

It is preferable therefore to use subtypes where possible or alternatively employ type conversion functions to match the actual type to the required type. These functions will not be synthesized into hardware as at the hardware level there is no notion of types or subtypes. These are discussed further in Chapter 6.

5.4.2 Understanding sequential signal assignments

Signal assignments can be performed in a concurrent or a sequential manner. For the designer, the difference between these two methods is that:

* concurrent assignments occur continuously so that the signal value is always up-to-date (neglecting propagation delays through components);
* sequential assignments do not take effect until a particular moment in time after they have been made inside a process.

Let us look at some of the traps it is easy to fall into when making sequential signal assignments. The following examples are simple but bad versions of the D-type flip flop. The function of each should be clear without showing the associated entity.

```
architecture BADI of DFF is
begin
   process
   begin
      wait until (CLK'event and CLK='0');
      Q <= D;
      QBAR <= not D;
   end process;
end BADI;
```

The architecture **BAD 1** assigns the flip flops' output values, Q and QBAR, inside the process. The rules for inferring flip flops state that any signal assigned inside a synchronous section becomes the output of a flip flop. This description therefore infers two flip flops as shown in Figure 5.32.

The architecture **BAD2** takes QBAR out of the process, ensuring that only one flip flop is inferred. However, in doing this, QBAR now becomes an unconditional concurrent signal assignment in which it is continuously assigned the inverted input D. Figure 5.33 illustrates this circuit. The solution is to replace 'not D' with 'not Q'.

```
architecture BAD2 of DFF is
begin
  process
  begin
    wait until (CLK'event and CLK='0');
    Q <= D;
  end process;
  QBAR <= not D;
end BAD2;
```

The architecture **BAD_JK**, uses two signals to store the present states of the flip flops' inverted and non-inverted outputs, SBAR and S. The **Case** statement inside the synchronous section is used to determine the new state of S and assign its value. The value of S is then passed on to SBAR at the end of the **Case** statement. Unfortunately, the new state is not registered on the flip flop inferred for S until the end of the process so the SBAR will be assigned the old state of S. If this is not bad enough, because SBAR is assigned inside the synchronous section, it also infers another flip flop!

```
architecture BAD_JK of JKFF is
signal S, SBAR : BIT;
begin
  process
  variable JK : BIT_VECTOR(0 to 1);
  begin
    wait until (CLK'event and CLK='0');
    JK := J & K;
    case JK is
      when "00" =>S <= S;
      when "10" =>S <= '1';
      when "01" =>S <= '0';
      when "11" =>S <= not S;
    end case;
    SBAR <= not S;
  end process;
  Q <= S;
  QBAR <= SBAR;
end BAD_JK;
```

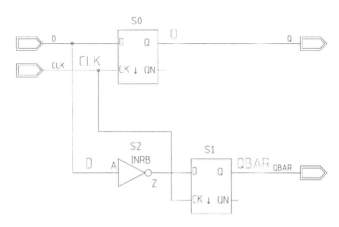

Figure 5.32 Synthesized circuit produced by BAD1.

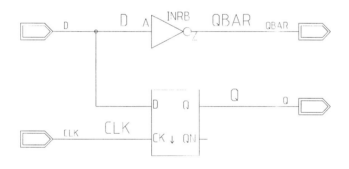

Figure 5.33 Synthesized circuit produced by BAD2.

5.4.3 Incomplete If statements

Case statements ensure complete coverage of all possible permutations of an expression by forcing any that are left undefined to be grouped together in an **Others** clause. The **If** statement does not force the designer to use a default **Else** and therefore allows the construction of incomplete **If** statements. The effect of an incomplete **If** statement has been demonstrated in this chapter.

Incomplete **If** statements can affect both signal and variable assignments. Consider the **If** statement:

```
if condition I then
    X <= ...
    Y := ...
end if;
```

The context of this statement will determine how the synthesizer deals with the incomplete If. If the statement is part of a synchronous section then it is equivalent to one of the following cases:

```
Y := 0;                          if condition I then
if condition I then                  X <= ...
    X <= ...                          Y := ...
    Y := ...                      else
else                                  X <= X;
    X <= X;                           Y := Y;
end if;                          end if;
```

In each of these two cases, the assignments made within the Else clause create feedback loops around the flip flops. This allows the present states of the object(s) to be preserved. In the left statement, the variable has been synthesized as a signal on a wire, not as the output of a storage element. On the other hand, in the right statement the variable is read before being assigned. It is therefore synthesized as the output of a flip flop with feedback.

- If the statement is part of a combinational section then two situations must be considered.
- If the signal (or variable) infers a storage element (in a synchronous section elsewhere) then the logic of the assignment becomes part of the asynchronous combinational input logic to the storage element.
- If the signal (or variable) does not infer a storage element the undefined state is considered to be a don't care.

5.4.4 Inferring level-triggered storage elements

The If and Wait statement templates demonstrated in this chapter are all used to infer edge-triggered storage elements. It is also possible to infer level-triggered elements – transparent latches. This can only be achieved with an If statement, again inside a process. However, the synthesizer does not use a clock expression template. The sequential block should take the form of the example shown below.

```
if (CLK='I') then
    ...
    REG <= X & REG(I to 3);
    -- other sequential statements
end if;
```

The important characteristic of this If statement is that REG (which could also be a variable) should be previously unassigned. When it is read in the assignment statement a value must be provided from somewhere and so storage elements are inferred. For the whole time that CLK is high, if X changes the value of REG must also be modified. Hence, the storage elements must be transparent latches.

This method of inferring storage elements relies on the fact that an unassigned signal or variable is being read. This is always the reason for the existence of unwanted sequential logic. If REG had a value before the assignment statement was encountered, purely combinational logic would be used in the implementation.

5.4.5 Understanding object types and literals

A hardware implementation can only represent values as groups of 1s and 0s. This is irrespective of the type – numeric, enumeration, etc. An integer value of 23, for example, will be represented by the binary pattern 10111. This does not mean that the designer can access individual bits of this string, as its VHDL type is INTEGER. To access these bits an object must be declared an appropriate type, such as BIT_VECTOR.

By the same token, it is not possible to treat a string of bits as, say, an integer and use it where a discrete value is required. For example, if S is a 4-bit BIT_VECTOR, the range specified in the following Case statement is illegal.

```
case S is
    when "1000" =>...
    when "1001" to "1100" =>...
    when others =>...
end case;
```

The synthesizer sees 1001 and 1100 as arrays of the enumerate type BIT, not as a direct and interchangeable representation of the discrete integer values 9 and 12. Note however that if S was declared as enumerate type CODES, which itself was defined as:

type CODES **is** (0000, 1000, 1001, 1111, 0110, 1100, 1010);

then each pattern is an enumeration literal representing a discrete encoded value. In this situation the Case statement shown below would be perfectly legal. Note that each pattern is actually a simple name not a string and therefore does not require double quotes.

```
case S is
    when 1000 =>...
    when 1001 to 1100 =>...
    when others =>...
end case;
```

5.5 EXERCISES

1 Design a counter that can count from 0 to 12. It counts the number of high values received on a 1-bit clock signal. Two versions need to be developed: the first counts up to 12 and on the 13th input returns to zero. The second counts to 12 and sticks at that value independent of the number of count signals received. If you have access to a logic synthesis tool compare the complexity of the two circuits.

2 Develop an up–down counter that can count up to 32. If the maximum number is reached a 1-bit output signal MAX is set high. If the counter reaches zero, the 1-bit output signal MIN is set high. In either case the counter will stop until the direction of the counter is changed.

3 Develop a barrel shifter. This is a widely used structure in microprocessor CPUs. This design is a shift register of 12 bits. It has a 1-bit input signal LEFT/RIGHT and a 3-bit input signal DISTANCE. LEFT/RIGHT identifies the direction in which the data in the shift register is moved. DISTANCE states the number of places that the data needs to be moved. At each end of the shift register is the LEFT_IO_PORT and the RIGHT_IO_PORT, both 1-bit bidirectional signals, which are used to load data into the registers.

5.5 SUMMARY

This chapter has introduced, discussed and demonstrated the principles of sequential logic design in VHDL. It has concentrated on the behavioural architecture style that is used to infer the presence of a section of sequential logic. A range of topics associated with arrays have also been discussed, including aggregate assignment and array operators. In summary, the following points should be remembered.

- Sequential logic can be instantiated but is usually inferred. The synthesizer will usually select D-type flip flops and use front-end combinational logic to create more complex sequential functions.
- A process containing either a Wait or an If statement must be used to infer sequential logic.
- A process that uses a Wait statement is fully synchronous.
- A process that uses an If statement can contain an asynchronous as well as a synchronous section.
- A method of initializing the sequential logic is absolutely vital. Default simulation values provided in signal or variable declarations are ignored by the synthesizer.
- Section 5.1.3 summarized the conditions under which sequential logic will be inferred when a clock template is present. These were as follows.

⤳ A signal assigned inside a synchronous section is synthesized as a flip flop. If it is assigned inside a combinational section it will be synthesized as a signal on a wire.

⤳ A variable that is used in the process *before* it is assigned a value is synthesized as a flip flop. Otherwise it is synthesized into a wire.

• Level-triggered storage elements (transparent latches) can be inferred using a level detection condition in an If statement, which then reads a previously unassigned signal or variable.

6

Arithmetic circuit design

6.1 BASIC PRINCIPLES OF VHDL ARITHMETIC

6.1.1 Which architectural style?

Arithmetic circuits are probably the hardest type to implement effectively using VHDL. It is easy to write an arithmetic assignment statement such as C<=A+B but (a) knowing how the synthesizer will model such a statement in hardware, and (b) whether one synthesizer will produce the same circuit as another, are issues that the designer must consider when choosing a design style. An adder circuit, for example, could be constructed using ripple-carry (RC) or carry-lookahead (CLA) techniques, but which one should be used and how can the designer indicate his or her preference to the synthesizer? Such questions are complicated further by the choice of target hardware.

Both the behavioural and dataflow-style architectures, give the designer the ability to specify *what* a circuit must do but not *how* it must do it. Each synthesizer will use its own operator templates that will create the same datapath logic block every time a particular operator is used. Resource sharing can be used to allow the same logic to be shared between several instances of an operator and hence reduce the circuit area. The designer can also exert some control over the operator's template by using overloading. The advantage of either of these styles (or a combination) is that technology independence is maintained in the actual code and the design can be ported easily to several different target technologies.

The structural style of creating an architecture allows the designer to specify explicitly:

- what type of adder or multiplier, for example, should be used;
- how each component/resource should be shared to make the most efficient use of area in a technology; and
- how operations should be scheduled to ensure that the critical timing path is minimized.

On top of this, vendors supply libraries that contain circuit modules or macros that are designed to make the most efficient use of the resources in a particular technology or actual device. These modules can be instantiated as components. The drawback of vendor libraries is that the VHDL description is technology dependent and to target a different technology can involve a considerable redesign effort. The macro/module approach is particularly valuable for FPGA devices that have available a limited number of fixed function resources.

6.1.2 Arithmetic operators

The arithmetic operators in VHDL can operate on any predefined numeric type. For a synthesizeable design their use is therefore limited to integer types. Box 6.1 illustrates ways of expressing synthesizeable numeric literals. The operators can be subdivided into the following classes.

- Adding operators (two operands) +   -
- Sign operators (one operand)   +   -
- Multiply operators   *   /   **mod   rem**
- Miscellaneous operators   **abs   ****

The concatenation operator **&** is also classed as an adding operator. This was covered in Chapter 5. The adding operators have the lowest precedence and the miscellaneous operators have the highest.

In order that the complexity of arithmetic hardware is constrained most synthesizers only allow power of 2 operations (2^M, where M is an integer) with the following operators.

- Division => $N/2^M$ (i.e. a shift right by M function)
- Exponentiation => $N**2$ (i.e. a squaring function, only!)
- Modulus => N **mod** 2^M
- Remainder => N **rem** 2^M

Often **Mod** and **Rem** are not supported at all for synthesis. None of these restrictions apply, however, when using the operators in a constant expression. Remember, though, that if the constant is a numeric then it must be an integer to synthesize.

The absolute operator **Abs** will return the magnitude of a numeric type or expression. Signed expressions and types are discussed below.

6.1.3 Signed arithmetic

In many cases arithmetic operations must be performed on signed values. These values may be stored as bit patterns in an array, in which case the signed shift functions **Sla** and **Sra** can be employed. Additionally, tool vendors will often provide an extension to the std_logic_1164 package that con-

Box 6.1 How to construct integer and based literals

The examples so far have illustrated the use of only standard integer literals such as 0, 10 and 255. However, like bit-string literals, numeric literals can be expressed in several forms. In a synthesis environment the use of numeric literals is limited to the integer and based types. Examples of the forms that integer literals may take are as follows.

   31 0   2E4   1E0   12_125

E indicates an exponent. The exponent value must be a non-negative integer if the literal is to be an integer; a negative value would produce a real literal, which cannot be synthesized. The underscore character can be used freely to aid the readability.

Based literals allow an integer value to be expressed in any base from 2 to 16. They have the general form:

   based_integer ::= base_integer # extended_digit {[underline] extended_digit}
                                              # [exponent]

The base integer specifies the base. The extended digits (0 to 9 and A to F) form the integer expression in that base and there is an optional exponent. The following examples illustrate based literals that are equivalent to the integer literal 128:

   2#1000_0000# 16#80# 16#8#E1 08#200#

Note that the exponent applies to the based literal not the equivalent integer literal. Also, depending on the base being used, only certain extended digits can be used. For example, base 2 can only use 0 or 1 in its expression.

   The value of these special forms of integer expression is to enable values to be represented in a way that is meaningful to the designer. They can all be treated as normal integers in any expression.

---

tains signed versions of all the appropriate standard operators for use with signed array types. These can be used by overloading the existing operator. Typically, a type will be defined as:

type SIGNED **is array (natural range <>) of** BIT;

When an operator such as + is used with two operands of this type the function will be called with the declaration:

function "+" (L: SIGNED; R: SIGNED) **return** SIGNED;

This will handle the operands as 2's complemented values and return an appropriate signed array. The rules governing this arithmetic are discussed in Box 6.2. Overloading and function declarations are discussed later in this chapter.

For integer arithmetic, the sign operators + and − can be used. These differ from the adding operators in that they take only one integer operand. It is generally advisable to enclose the sign and the operand in parentheses to clarify the meaning of an expression. It is also illegal to follow one operator with another. For example:

```
A <= B/(−C);      -- legal
A <= B/−C;        -- illegal (and confusing!)
```

Remember that these sign operators can be used with numeric types only (therefore only integers when synthesizing). The following expression is illegal.

signal D : BIT_VECTOR(7 **downto** 0);

...

D <= − "1001_0000"; -- negation operator only defined for numerics

Box 6.2 How to handle array arithmetic

Tool vendors often provide arithmetic functions for operations on unsigned and signed array types. Signed array arithmetic operations must be handled differently to unsigned. Vendors usually provide unsigned array arithmetic operators for BIT_VECTORs, X01Z_VECTORs (Chapter 7) and declare a new type such as UN-SIGNED. Signed array arithmetic operators are often provide for STD_LOGIC_VECTORs (Chapter 7) and a new type such as SIGNED. 2's complement notation is generally used for signed array types. The following rules are applied in array arithmetic.

+, −	Operands can be different sizes. The result is the same length as the longest operand.
*	Operands can be different sizes. The result is the combined lengths of the operands for unsigned types and is 1 bit shorter for signed types.
/	Operands must usually be <32 bits but can be different sizes. The result is the same length as the left operand.
rem	Operands must usually be <32 bits but can be different sizes. The result is the same length as the right operand.
abs	Result is the same length as the operand.

6.2 ADDER CIRCUITS

This section is structured differently to sections 4.2 and 5.2. Its aim, however, is still the same – to introduce and demonstrate the use of new VHDL constructs and explain how these are transformed into hardware by synthesis. The aim is not to teach digital logic design by presenting a wide range of different addition algorithms and then comparing their characteristics in hardware. This will only be done to illustrate the synthesis process and to show how different design approaches can influence the performance and area characteristics of a circuit.

Not all the architectural descriptions presented in this section are synthesized and/or optimized. One particular circuit contains no new constructs but is a useful example for demonstrating the process of speed optimization.

The circuits designed below are primarily adders; however, they can be modified to provide a subtraction capability as well. Some examples of these are also demonstrated.

The constructs introduced in this section include:

- built-in and user-defined functions;
- operator and subprogram overloading;
- qualified expressions;
- sequential statements
    For-loop
    While-loop
    Loop (forever)

6.2.1 Functional overview

Three different forms of addition (and subtraction) circuits are demonstrated below.

- *Integer.* These use the built-in integer arithmetic operators + and – to show how the synthesis tool implements standard operators.
- *Ripple carry.* These are user-defined versions of the built-in operators that demonstrate how a flexible arithmetic circuit can be constructed and called in different ways.
- *Carry lookahead.* This is again user defined and demonstrates how a relatively complex algorithm can be implemented using a few components and a hierarchical structure.

In each case, the addition (or subtraction) is of two 4-bit signed numbers giving a signed 4-bit result. In some circuits a carry out or overflow bit is implemented.

The synthesizer implements the integer arithmetic functions in its predetermined way. As will be seen, this is a sign-magnitude format. The user-defined circuits, however, all use an array type called **SIGNED** with elements of type **BIT**. This represents a 2's complemented signed number. Hence, the architectural descriptions produced for the user-defined adder units can be (and are) easily converted into subtraction units.

6.2.2 VHDL code design

This section looks at the complete design of each style of arithmetic circuit in turn rather than discussing the libraries, entity and then architectural descriptions. It is therefore organized by design rather than design unit as before.

Integer arithmetic

The addition and subtraction operators are defined in the **STANDARD** package for the type **INTEGER**. Implementing integer arithmetic is therefore very straightforward and the code is deceptively compact – arithmetic operators can produce very large amounts of logic when implemented in hardware.

To recap on Chapter 3, integers are implemented in a bitwise format by the synthesizer but it is not possible for the designer to address individual bits of the integer. A negative integer is stored with sufficient bits to represent the magnitude of the integer plus an additional sign bit – sign-magnitude format.

The simplest version of an adder circuit is shown in Figure 6.1. This description does not introduce any new language elements but it is useful for later comparisons. Note that the input and output signals have predefined signed integer ranges. It is not usually sufficient to define the input ranges and let the outputs determine appropriate upper and lower bounds themselves. If ANSWER was left unconstrained it may use a 32-bit representation. This is synthesizer dependent and redundant outputs will be removed by area optimization.

```
entity INTGR_ADDER is
    port (   W, X : in INTEGER range -8 to 7;
             ANSWER : out INTEGER range -8 to 7
             );
end INTGR_ADDER;

architecture ONEADD of INTGR_ADDER is
begin
    ANSWER <= W + X;
end ONEADD;
```

Figure 6.1 Integer adder using the built-in operator.

The second integer arithmetic circuit implements both an addition and a subtraction operation. As Figure 6.2 shows, the MODE input determines whether X is added to or subtracted from W. A Selected signal assignment statement is used to implement the behaviour. The result in either case is still a signed integer. Again, no new constructs are introduced but the important feature of this circuit is the presence of both the addition and subtraction operators. As MODE is an external signal both the operators must be synthesized as the circuit will not know which operation to perform until MODE is known. This point is noted as examples in the ripple-carry arithmetic section take a different design approach, which reduces the logic requirements.

It is interesting to observe how this architecture is translated into hardware and whether the synthesizer can share any common logic structures between the operators' implementations. The circuit produced will be used in comparisons with those of the user-defined arithmetic circuit descriptions.

Ripple-carry arithmetic

It is generally more useful and makes a design clearer if adders and other arithmetic functions are designed to operate on bit-oriented types. Many synthesis tool vendors will usually provide sufficient numbers of different bit array types such as UNSIGNED and SIGNED, not to mention BIT_VECTOR along with appropriate arithmetic operations. The SIGNED type is concentrated on in these examples and is declared as a new type. This frees the reader from any tool dependency when studying these examples. As discussed in section 6.1.3, the SIGNED type represents a 2's complemented binary number. To represent the integer range −8 to 7 therefore requires 4 bits.

```
entity INTGR_ADDSUB is
    port (   W, X : in INTEGER range -8 to 7;
             MODE : in INTEGER range 0 to 1;
             ANSWER : out INTEGER range -8 to 7 );
end INTGR_ADDSUB;

architecture DATAFLOW of INTGR_ADDSUB is
begin
    with MODE select
        ANSWER <=  W + X   when 0,
                   W − X   when 1;
end DATAFLOW;
```

Figure 6.2 Integer addition and subtraction unit using the built-in operators.

Remember that the aim of these examples is to demonstrate a number of new VHDL constructs, not to investigate the merits and problems of ripple–carry arithmetic.

There are a variety of approaches that could be adopted to implement a user-defined routine such as a ripple-carry adder.

- Construct a component that can be instantiated in a design when required
  Like the INTGR_ADDSUB example in Figure 6.2, the whole architectural description would be synthesized as the compiler does not know which features or functions within the component will be used by a particular design. This is therefore a bad approach if the component must implement a flexible 'selectable' function.

- Place it inside a process
  This is fine if the routine is only used in this particular design but a library stored version would be more practical if there is a possibility of future reuse. It will also make the code unnecessarily long and more difficult to follow.

- Store it in a package as a subprogram.
  Like other programming languages, the two types of subprogram available in VHDL are functions and procedures. The built-in arithmetic operators all call functions and so these will be used in this case. Procedures are more flexible than functions and are discussed further in Chapter 7.

The package is shown in Figure 6.3 and contains the declaration for the new type SIGNED and four function declarations. A function can receive any number of parameters (signals, variables and constants) in its parameter list and returns a single result of a specified type. Parameter lists were discussed in detail in Chapter 5. The function is constructed in two parts. Its declaration is placed in the package as shown and the actual functional description – the function body – is located in the package body. The package body associated with this package is shown in Figure 6.4.

```
package ARITH_TYPES is

    type SIGNED is array(NATURAL range <>) of BIT;
    function RIPPLE(A, B : SIGNED) return SIGNED;
    function RIPPLE(A, B : SIGNED; C : BIT) return SIGNED;
    function "+"(A, B : SIGNED) return SIGNED;
    function "-"(A, B : SIGNED) return SIGNED;

end ARITH_TYPES;
```

Figure 6.3. The package ARITH_TYPES containing the new arithmetic routines.

A function is flexible. As the package body of ARITH_TYPES shows, the parameters passed and returned from a function do not have to be constrained. If the function is designed without constraints it can be used in many different designs or several times in the same one, performing operations on n-bit objects. To achieve this flexibility, attributes are used to provide information about each object passed to the function. The RIPPLE functions use the **Left** and **Reverse_range** attributes discussed further in Box 6.3. A component, on the other hand, is constructed from an entity declaration and an architectural body. This means that the size of each I/O path must be determined at the component's compile time, reducing its flexibility.

A function is very similar to a Process statement in its operation except that it cannot contain signal assignment or Wait statements. This means that a function cannot be used to create sequential logic. As the RIPPLE functions illustrate, variables that are local to the function can be declared in the declarative part and its execution is sequential. All objects declared locally are discarded when the final result is returned. The syntax of a function and other important details are given in Box 6.4.

The package contains two functions called RIPPLE, both of which implement ripple-carry arithmetic. The reason for the two functions having the same name is discussed shortly. The first function adds two 2's complemented signed numbers and returns not only the result but also an overflow bit. As a function can only return one value, the overflow must be concatenated with the result. Hence, the addition of two 4-bit numbers

```
package body ARITH_TYPES is

-- Ripple carry addition, returns overflow+2's
-- complemented result
function RIPPLE(A, B : SIGNED) return SIGNED is
    variable CARRY : BIT_VECTOR(A'left+1 downto 0);
    variable SUM : SIGNED(A'left+1 downto 0);
begin
    SUM := (others => '0');
    CARRY := (others => '0');
    -- reverse the range of A so that counting begins at bit 0
    for K in A'reverse_range loop
        SUM(K) := A(K) xor B(K) xor CARRY(K);
        CARRY(K+1) :=  (A(K) or B(K)) and
                (A(K) or CARRY(K)) and
                (B(K) or CARRY(K));
    end loop;
```

-- continued on next page

Figure 6.4 The package body containing the new arithmetic routines.

```
    -- check for overflow and set overflow (carry out) bit in the result
    if CARRY(A'left+1) /= CARRY(A'left) then
        SUM(A'left+1) := '1';
    end if;
    return SUM;
end RIPPLE;
```

--Ripple carry addition/subtraction,returns 2's
--complemented result,no overflow
```
function RIPPLE(A, B : SIGNED; C : BIT) return SIGNED is
    variable CARRY : BIT_VECTOR(A'left+1 downto 0);
    variable VB : SIGNED(B'range);
    variable SUM : SIGNED(A'left downto 0);
begin
    SUM := (others => '0');
    CARRY := (0 => C, others => '0'); -- the carry in selects add or
subtract
    VB := B;
    if C='1' then
        VB := not B;          -- invert B for subtraction
    end if;
    for K in A'reverse_range loop
        SUM(K) := A(K) xor VB(K) xor CARRY(K);
        CARRY(K+1) :=  (A(K) or VB(K)) and
                       (A(K) or CARRY(K)) and
                       (VB(K) or CARRY(K));
    end loop;
    return SUM;
end RIPPLE;
```

-- Addition of two signed numbers
-- Uses the ripple carry adder through the '+' operator.
```
function "+"(A, B : SIGNED) return SIGNED is
begin
    return RIPPLE(A,B);
end;
```
-- Subtraction of two signed numbers.
-- Uses the ripple carry adder/subtractor through
-- the '-' operator
```
function "-"(A, B : SIGNED) return SIGNED is
begin
    return RIPPLE(A,B,'1');
end;
end ARITH_TYPES;
```

Figure 6.4 (Continued from previous page.)

Box 6.3 How to use predefined attributes

The characteristics of a signal, an array or a type in a description are stored by the **attributes** of that entity. An attribute provides the designer with additional useful information that allows the behaviour to be specified more clearly and simply, and enables the description to be generalized. A number of predefined attributes exist in VHDL but only a few are supported for synthesis. Some, such as Event and Stable are only supported when used in specific ways. The following attributes are generally supported.

Type attributes

T'base Returns the base type of the type or subtype T. This must be used only as a prefix to another attribute.
T'left Returns the left bound value of the scalar type or subtype T.
T'right Returns the right bound value of the scalar type or subtype T.
T'high Returns the high bound value of the scalar type or subtype T.
T'low   Returns the low bound value of the scalar type or subtype T.

Examples:

> **subtype** T **is** INTEGER **range** 10 **downto** 3;
> **type** PRIMARY **is** (red, blue green);
>> T'**left** is equivalent to 10
>> T'**low** is equivalent to 3
>> PRIMARY'**base'right** is equivalent to green

Array attributes

The left, right, high and low attributes can also be applied to an array object or a constrained array subtype. The following can also be used with the these objects or subtypes:

A'range	Returns the discrete range of A.
A'reverse_range	Returns the discrete range of A in a 'last to first' format.
A'length	Returns an integer expression indicating the number of elements in A.

Each of these attributes may take a postfixed expression, N, where N is a static integer expression indicating the index range of A to which the attribute applies. This is useful for two-dimensional arrays.

Examples:

> **type** A **is array** (0 **to** 15, 0 **to** 3) **of** BIT;
>> A'**range**(1) is equivalent to 0 **to** 15
>> A'**length**(2) is equivalent to 4
>> A'**high**(1) is equivalent to 15

-- continued

Signal attributes

S'stable   Only returns TRUE if the signal S is stable at this instant in time.

S'event   Only returns TRUE if the signal S changes state at this instant in time.

These attributes are usually only supported in expressions that specify events occurring on clocks. As such, they can only be used in a Wait or If statement inside a process to indicate the presence of a sequential section.

---

will return a 5-bit result. This is a limitation of a function but it can usually be circumvented.

The function implements the ripple-carry addition process in a bitwise manner. It is assumed throughout these examples that both the inputs, A and B, are of the same length. A new statement, the For Loop, is used to control the addition process for each bit of A and B. This statement is similar to the Loop statement in other programming languages. Again attributes are useful for determining the number of iterations that are made around the loop. Here **Reverse_range** inverts the descending index range of A to produce an ascending counter that uses K as the index parameter. This is implicitly declared in the For-Loop statement. The designer will generally require a For Loop to execute for all the specified number of iterations. It is possible, however, to terminate a single iteration or even the whole loop if required at any point. This is usually only needed in loops without a fixed number of iterations, such as a While-Loop or an Infinite Loop. These types of loop and their control are discussed in section 6.3.2, where they are used in a multiplier example. These are all sequential statements, unlike the For-Generate, which was concurrent. The syntax and other key points about the For-Loop are given in Box 6.5.

The second RIPPLE function is similar to the first but also allows subtraction. As 2's complemented numbers are being used, changing addition into subtraction is only a matter of inverting the bits of B and setting the carry in for the least significant bit. To determine which function is required, this carry in is provided as an input to the function through its parameter list. Initialization of the carry bit array inside the function therefore uses an aggregate expression with named association. This places the carry in C in the least significant bit and resets all others. VB stores the appropriate form of the input B – inverted or non-inverted. This is used in the For-Loop instead of B. A carry out is not returned by this function; therefore, 4-bit inputs will return a 4-bit result.

Box 6.4 How to declare a function

A function is a subprogram that can receive a number of parameters and return a result of a specified type. It can consist of one or two parts. The **function declaration** defines the function's name, its parameters and return value type. It has the following syntax.

function_declaration ::= **function** simple_name [(parameter_list)] **return**
                                                                    type_mark ;

The parameter list is a form of interface list (Chapter 5) that can consist of constants, variables and signals of mode **In**. This mode is the default and does not need to be specified. When the object type is not specified, a constant is assumed. Objects may be declared as unconstrained array types, allowing their length to be determined each time the function is called. Attributes such as LENGTH and RANGE can then be used inside the function to achieve this. The **function body** contains the description of the behaviour of the function. It has the following syntax.

function_body ::=
    function simple_name [(parameter_list)] **return** type_mark **is**
        function_declarative_part
    begin
        sequential_statements
        return expression;
    end [**function**] [simple_name];

1 The function declarative part may contain declarations for subprograms, types, subtypes, constants, variables, attributes. It may also contain subprogram bodies, attribute specifications and use clauses. Several other declarations are not supported for synthesis. Every declaration made in this part is reset between calls. Initial values for objects will be ignored when synthesizing.

2 Any sequential statements can be used in a function *except* a Signal assignment and a Wait. This means that a function cannot infer a sequential block of logic.

3 Any number of Return statements can occur in the function body. An expression of the type or subtype specified by the type mark must be returned.

4 Functions are usually stored in packages. The function declaration is stored in the package declaration and the function body in the package body. A declaration is not required for the function if its body is located in the declarative part of an entity, architecture, procedure or another function. The visibility of the function is however restricted in any of these cases.

Box 6.5 How to create For-Loops

1 A For-Loop statement enables a set of other sequential statements to be repeated for a fixed number of iterations. It has the following syntax.

for_loop_statement ::=
[loop_label :]**for** iteration_parameter **in** value1 direction value 2 **loop**
      sequential_statements
end loop [loop_label];

2 To synthesize a For-Loop, value1 and value2 must be static, implying a fixed number of iterations around the loop. Theses values do not necessarily need to form a numeric (integer) range; any discrete type, such as an enumerate, may be used. For example, a range could be defined as 'INDIGO **downto** ORANGE' given the type:

type RAINBOW **is** (RED, ORANGE, YELLOW,
        GREEN, BLUE, INDIGO, VIOLET);

3 The attributes **Range**, **High** and **Low** are also often used to define the range of the For-Loop from some other parameter. This parameter must be a constant to ensure that the range is static.

4 For an ascending range direction is **To** and for a descending range direction is **Downto**.

5 Any sequential statements can be used inside the loop *except* a Wait statement. Therefore a For Loop can only be used to create combinational logic. This is constructed by unfolding the loop during compilation.

6 The iteration parameter cannot be modified inside the For Loop. It must be treated as a constant that has been implicitly declared.

7 By acting on the present value of the iteration parameter or any other condition, the present iteration of the loop or the whole For Loop statement can be terminated. See Box 6.12

8 A nested loop must be fully contained within its outer loop. It is advisable to use loop labels for identification when nesting loops.

9 Another type of For-Loop is For-Generate, discussed in Chapter 5. See also Box 6.11.

The two other functions in the package show how new operators can be created for user-defined types. These highlight some important features of functions, operators and a technique known as overloading.

- A function can only be called as part of an expression in either a sequential or concurrent statement. In the functions + and – the RIPPLE functions return a result that itself is immediately returned to the calling architecture. The calling of functions in general is discussed in Box 6.6.
- Operators and functions can be **overloaded**. This involves the substitution of one version of an operator or function for a different one with an identical name. RIPPLE is an example of an overloaded function where the + operator function overloads the built-in + operator functions.

Look at the example architectures that call this package's functions in Figures 6.5–6.7. The architecture RIPPLE1 calls the function RIPPLE in a signal assignment expression. How does the synthesizer know which version of RIPPLE is being called? In this case the number of parameters in the parameter list reveals which version is required. As there are only two, it is the first version. In this example, therefore, the first RIPPLE function has overloaded the second owing to the parameter count. This function returns an overflow bit attached to the result. This is extracted and the outputs, ANSWER and OVERFLOW are assigned.

Box 6.6 How to call a function

A function is called as part of an expression inside a concurrent or sequential statement. When synthesizing a function it cannot be called recursively. In the example below the conversion function 'To_Integer()' may return either an integer or any integer subtype that can be assigned to B.

```
signal A : BIT_VECTOR(7 downto 0);
signal B : INTEGER;
...
A <= "11001000";
B <= 10 + To_Integer(A);      -- B equals 10+200
```

This is an example of a type conversion function, which is covered in Chapter 7. Such a function can also be called *inside* the parameter list if there is a mismatch between one of the parameters being passed and the type the function actually requires. This usage is allowed because type conversion functions are **built in** (Box 6.10).

A parameter list can be constructed using both positional and named association. The latter is simply a clearer and less error-prone way of passing the parameters to a function. Parameter lists and interface lists in general were discussed in Chapter 5.

```
library ARITH;
use ARITH.ARITH_TYPES.all;
entity ADD I is
   port (    A, B : in SIGNED(3 downto 0);
             OVERFLOW : out BIT;
             ANSWER : out SIGNED(3 downto 0)
             );
end ADD I ;

architecture RIPPLE I of ADD I is
signal RESULT : SIGNED(4 downto 0);
begin
   RESULT <= RIPPLE(A,B);
   ANSWER <= RESULT(3 downto 0);
   OVERFLOW <= RESULT(4);
end RIPPLE I ;
```

Figure 6.5 Calling the signed binary RIPPLE addition function.

```
library ARITH; use ARITH.ARITH_TYPES.SIGNED;
entity ADD2 is
   port (    A, B : in SIGNED(3 downto 0);
             OVERFLOW : out BIT; ANSWER : out SIGNED(3 downto 0));
   -- see Box 6.7 for more information on this attribute
   attribute BUILT_IN_OPERATORS : BOOLEAN;
   attribute BUILT_IN_OPERATORS of ADD2 : entity is FALSE;
end ADD2;
architecture RIPPLE2 of ADD2 is
signal RESULT : SIGNED(4 downto 0);
begin
   -- call operator function with a Selected name
   RESULT <= ARITH.ARITH_TYPES."+"(A,B);
   ANSWER <= RESULT(3 downto 0);
   OVERFLOW <= RESULT(4);
end RIPPLE2;
```

Figure 6.6 Calling the addition operator function + for signed binary numbers.

The RIPPLE2 architecture calls the same function in an indirect way through the addition operator function. This overloads all the built-in addition operator functions that are provided for INTEGER and non-synthesizeable types including REAL. The attribute BUILT_IN_OPERATORS discussed shortly in Box 6.7 is set to FALSE so that this occurs. When overloading an operator only two parameters must be supplied (the sign op-

```
library ARITH; use ARITH.ARITH_TYPES.SIGNED;
entity ADD3 is
    port (    A, B : in ARITH.ARITH_TYPES.SIGNED(3 downto 0);
              ANSWER : out ARITH.ARITH_TYPES.SIGNED(3 downto 0));
end ADD3;

architecture RIPPLE3 of ADD3 is
begin
    -- no overflow for this adder/subtractor
    ANSWER <= ARITH.ARITH_TYPES.RIPPLE(A,B,'0');
end RIPPLE3;
```

Figure 6.7 Calling the signed binary addition/subtraction RIPPLE function.

erator must only have one). This overloaded operator is called using function notation in which the operator symbol appears in quotes before its parameters. The standard operator notation can also be used.

This architecture also illustrates the use of Selected names to access libraries and packages. The ARITH library contains the compiled package ARITH_TYPES and to access the SIGNED type declaration the Library and Use clauses have been used in the normal way. To access the "+" operator function a Selected name has been used. This directly selects the library, package and function without the need for a Use clause. Section 3.4.3 discussed accessing package items in this way.

The third architecture RIPPLE3 takes this direct access a step further and dispenses with the Use clause altogether. Both the SIGNED type declaration and the RIPPLE function are accessed directly. This architecture accesses the second RIPPLE function, overloading the first by the number of parameters that are supplied. This is only one way of achieving function overloading; Box 6.8 discusses these methods further. Another approach selects the appropriate function by the type of the parameters that are supplied.

Operator overloading uses this second approach, as the number of parameters in the original or overloading operator function is always the same. In some cases, this can lead to ambiguous calls in which the synthesizer cannot determine the required function. Consider the architecture RIPPLE4 in Figure 6.8. This calls the second RIPPLE function and performs the subtraction operation. When the + operator function was called in RIPPLE2 a selected name was used to specify explicitly the location of the operator's function. In this case, the synthesizer will search all visible packages for a – function that takes two parameters of type SIGNED. If there is more than one then a selected name should be used to indicate which should be used. See Box 6.7 for more general details on this topic.

Another problem will arise if different versions of the operator function exist for enumeration types which contain literals that are identical. The

Box 6.7 How to overload an operator

Operator overloading uses the same principle as subprogram over-loading. Any of the predefined language operators (arithmetic, logical, relational, etc.) can be overloaded. Examples of arithmetic operator overloading for signed bit array types have already been shown in section 6.1.3. Vendors provide these and other operators to enable more object types to be supported and make descriptions clearer. The STD_LOGIC_1164 package contains overloaded logical operators for the STD_LOGIC_VECTOR type (Chapter 7). Vendors will supply over-loading functions for BIT_VECTORs, which may take the form:

function "+" (L, R : BIT_VECTOR) **return** BIT_VECTOR;

Whenever the + operator is encountered in a description and the operands are both of type BIT_VECTOR, the overloaded function will be used. Note that all operators require two operands except the sign operators, which take one. An operator cannot be overloaded with a procedure. When using an overloaded operator it can be referenced in two ways – with operator or function notation – as shown below.

```
W <= X + Y;      -- operator notation
W <= "+" (X,Y);  -- function notation
```

Replacing existing operators by overloading

The above shows how to **extend** the capabilities of VHDL by over-loading. In this case, the original operator is still available and can be called by providing parameters of the appropriate type. But what happens if you actually wish to **replace** the present behavioural definition of an operator with a new one? This may be required if:

- the designer wants to ensure that an adder, for example, is implemented in hardware as a carry lookahead and not a ripple carry; the new description of the adder's behaviour would be explicit enough to ensure that correct hardware was constructed;
- the designer wishes to use a particular technology that provides its own libraries of hardware-efficient implementations of specific operators; these are often in the form of macros that can be used to build N-bit blocks of datapath logic.

A synthesizer therefore needs to give the designer the ability to over-ride the built-in functions and operators. This is often done through an attribute declaration in the entity, such as:

attribute BUILT_IN_OPERATORS of entity_name : **entity is** FALSE;

This switches off the use of internal operators and forces the synthesizer to search externally.

```
library ARITH; use ARITH.ARITH_TYPES.all;
entity SUBTRACT is
    port ( A, B : in SIGNED(3 downto 0);
            ANSWER : out SIGNED(3 downto 0) );
    attribute BUILT_IN_OPERATORS : BOOLEAN;
    attribute BUILT_IN_OPERATORS of SUBTRACT : entity is FALSE;
end SUBTRACT;

architecture RIPPLE4 of SUBTRACT is
begin
    ANSWER <= "-"(A,B);
end RIPPLE4;
```

Figure 6.8 Calling a signed subtraction function.

same problem will occur for array types that use the same element types. For example a BIT_VECTOR and a SIGNED type both contain BIT elements and could both make the following assignment:

```
ANSWER <= "-"( "10011100", "00101101" );
```

Assuming that – operator functions are defined for both array types and are located in a visible library, how does the synthesizer know whether these bit-string literals represent BIT_VECTORs or SIGNED binary numbers? The designer must therefore qualify the type of each literal, namely:

```
ANSWER <= "-"( SIGNED'"10011100", SIGNED'"00101101" );
```

Solving the problems of ambiguities such as this in VHDL and using qualified expressions is discussed in Box 6.9.

Finally, in this section is the addition and subtraction unit that selects either operation through the external MODE input. This is shown in Figure 6.9 and is included to allow a comparison of its circuit with that of the integer addition/subtraction function, INTGR_ADDSUB-DATAFLOW shown in Figure 6.2. As in this earlier version, if MODE is low then X is added to W; if MODE is high it is subtracted. By using only one call to a single function that integrates both arithmetic operations, it is hoped that logic can be more easily and efficiently shared between them. This design approach therefore effectively implements a manually controlled resource allocation operation.

Carry-lookahead arithmetic

Using built-in operators to implement arithmetic processes is fine if the designer is not too concerned about the area and speed characteristics of the hardware that is produced; optimization will usually enable constraints to be met. The trouble with this is that what you get is not always what you

Box 6.8 How to overload a subprogram

Overloading allows a user to:

- define a new behaviour or structure for a subprogram or operator; and
- describe the behaviour of an operator or subprogram for a new type of object.

VHDL is object oriented and therefore operators and subprograms are treated as objects to which a certain number of parameters can be supplied and returned. If two subprograms with the same name exist but the number, name and/or type of parameters are different then one object is said to **overload** the other. An operator is overloaded when a new version exists that takes the same number of operands but of a different type. A package containing four declarations of a subprogram named PARITY is shown below.

```
Package PARITY_FUNCTIONS is
function PARITY ( DATA_A : BIT_VECTOR(0 to 7); CHK : BIT) return
                                                     BOOLEAN;
function PARITY ( DATA_A : X0IZ_VECTOR(0 to 3); CHK : BIT) return
                                                     BOOLEAN;
function PARITY ( D1, D2 : BIT_VECTOR(0 to 7); CHK : BIT) return
                                                     BOOLEAN;
function PARITY ( D1, D2 : BIT_VECTOR(0 to 7); CHK : BIT) return BIT;
end package PARITY_FUNCTIONS;
```

The second function overloads the first by changing the **type** (or subtype in this case) of the arguments in the parameter list. The third overloads the first by changing the **number** of arguments in the parameter list and it overloads the second by changing both the type and number of arguments. Calls to each function are illustrated below.

```
signal A,B : BIT_VECTOR(0 to 7);
signal C : X0IZ_VECTOR(0 to 3);
signal Y,Z : BIT;
variable PARITY_CHK : BOOLEAN;
PARITY_CHK := PARITY(A, Z);      -- calls 1st function, 'original'
PARITY_CHK := PARITY(C, Z); -- calls 2nd function, type of parameter changed
PARITY_CHK := PARITY(A, B, Z);-- calls 3rd function, no. of parameters changed
Y <= PARITY(A, B, '0')      -- calls 4th function, return value must be a BIT type
```

```
library ARITH; use ARITH.ARITH_TYPES.all;
entity SIGNED_ADDSUB is
    port (   W, X : in SIGNED(3 downto 0);
             MODE : in BIT;
             ANSWER : out SIGNED(3 downto 0)  );
end SIGNED_ADDSUB;

architecture DATAFLOW of SIGNED_ADDSUB is
begin
        ANSWER <= RIPPLE(W,X,MODE);
end DATAFLOW;
```

Figure 6.8 Implementing an addition/subtraction function with RIPPLE.

want, particularly when faced with relatively rigid hardware resources in technologies such as FPGA. Structuring the design more carefully will allow synthesis and optimization of each element to be carried out in a more controlled yet still automated manner. Also, if a clearly defined algorithm exists for a process, why not use it?

The carry-lookahead adder is designed to speed up the addition process by overlapping the carry and sum operations. The result is that the $2n$ logic level ripple-carry arithmetic process is reduced (in practice) to a $\log_2 n$ logic level operation.

It is not the intention here to determine the relative merits of carry-lookahead techniques over ripple carry. Instead, this section aims to demonstrate how an algorithm of this type can be implemented in VHDL. A highly structured approach has been adopted in the design, and the benefits of this will be illustrated when the design is synthesized. No new language constructs are introduced here.

The full details of the carry-lookahead algorithm are not of particular interest here and so will be covered in only the minimum of detail. The adder to be designed performs 2's complement addition on two 4-bit signed binary numbers, producing a 4-bit result with overflow detection. Unlike the ripple-carry operations described in the previous section, this adder is constructed as a component rather than a function. This means that it is fully synthesized, producing a design database of components before being instantiated in another design. Therefore, unlike a function, all the parameters of the adder, such as its bit width, must be constant. This obviously limits the flexibility of the adder and its ability to share logic resources with other elements in the design.

A 1-bit full adder is formed using the *SUM* and carry out *COUT* equations:

$$SUM = A \ xor \ B \ xor \ CIN \tag{6.1}$$

$$COUT = (A \ and \ B)or(A \ and \ CIN)or(B \ and \ CIN) \tag{6.2}$$

Box 6.9 How to handle ambiguous statements

An ambiguous statement can arise if it is not possible to determine which enumeration type is being used. One consequence of this is that the synthesizer may not be able to determine which version of a subprogram is being called. This often occurs when a function or operator has been overloaded. Consider the package PARITY_FUNCTIONS_2 shown below.

```
package PARITY_FUNCTIONS_2 is
function PARITY ( DATA_A : BIT_VECTOR(0 to 7); CHK : BIT) return
                                                      BOOLEAN;
function PARITY ( DATA_B : BIT_VECTOR(0 to 7); CHK : BIT) return
                                                      BOOLEAN;
function PARITY ( DATA_B : X0IZ_VECTOR(0 to 7); CHK : BIT) return
                                                      BOOLEAN;

end package PARITY_FUNCTIONS_2;
```

Using the objects declared in Box 6.8, the call

```
PARITY_CHK := PARITY( A, Z );
```

is ambiguous because it is not clear whether the first or second version of the function is required. This can be avoided by using named association in the parameter list, namely

```
PARITY_CHK := PARITY( DATA_B=>A, CHK=>Z );
```

This is clearly a call to the second function. Now consider the following call.

```
PARITY_CHK := PARITY("01001111", Z );
```

This could be a call to either the second or the third function because the type of the vector 01001111 is not clear – it could be either BIT_VECTOR or X0IZ_VECTOR (Chapter 7). The expression can be qualified to make the choice clear, i.e:

```
PARITY_CHK := PARITY( X0IZ_VECTOR'("01001111"), Z );
```

A **qualified expression** is one in which the type is explicitly stated and can be used whenever the type of an expression is not clear from its context. It takes the form:

```
qualified_expression ::= type_mark'( expression )
```

As the above example demonstrates, ambiguous expressions that need qualifying usually occur because the same enumeration literal(s) occurs in more than one type declaration.

A carry-lookahead adder introduces the **propagate** term P and the **generate** term G such that the carry out equation can be rewritten as:

$$COUT = G \text{ or } (P \text{ and } CIN)$$ (6.3)

where G and P are defined as shown in Equations 6.4 and 6.5. These equations together can then form the well-known structure of a carry-lookahead adder shown in Figure 6.10. As this shows, the carry in is fed up through the circuit at

$$G = A \text{ and } B$$ (6.4)

$$P = A \text{ or } B$$ (6.5)

the same time as the bits to be added are fed in from the top. The G and P terms that are generated by inputs A and B, flow down through the circuit and combine with the carry bits to determine the final carry out. This is generated by a unique cell, C. As this diagram illustrates, there are three different components that can be replicated as required to implement the algorithm. The entities and architectures for the cells A, B and C are shown in Figure 6.11. Rather than repeating the Boolean equations that each cell implements it is just as easy to refer to the architecture bodies.

Cell A forms the top level of the calculation for each pair of input bits, generating the associated G and P values. These are propagated down to the B cells where the intermediate-level logic calculates each carry term or G and P values for the lower stage. When a B cell has generated a carry it is passed to the A cell to which it belongs. In this case, as C_2 is generated at the third level of the tree it is also passed back to the next layer of B cells to enable the next carry bit, C_3, to be generated correctly. The tree structure can be extended or reduced in many ways, affecting the performance and area characteristics of the circuit. By increasing the functionality of each cell the tree would be extended and a reduced functionality would cause it to compact. The structure used here is commonly employed. The generic components in Figure 6.12 help to determine how the generate, propagate and carry signals in Figure 6.10 map into the port lists of the components CLA_CELL_A, CLA_CELL_B and CLA_CELL_C.

One of the benefits of using a modular approach is its expansion capabilities. If a behavioural-style architecture is used it is possible to increase the bit width of the adder by changing the ranges of arrays. Unfortunately, this would then require a long resynthesis and optimization process involving the whole design. A modular approach means that any number of small, presynthesized and optimized components can be instantiated in a top-level description without causing a noticeable increase in processing time. The synthesis process itself is only a matter of generating the interconnect for a few more components. The top-level architecture that binds all the cells in this design together is shown in Figure 6.13.

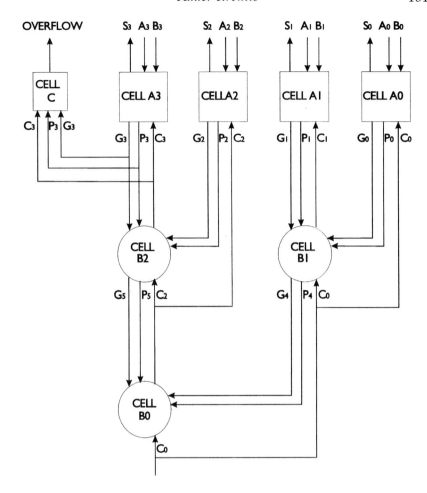

Figure 6.10 Cell structure of a 4-bit carry-lookahead adder with overflow.

A named association of input signals to component ports has been used in all the components in this design due to the number of similar names. The A type cells have a very regular structure and so a regular **Generate** statement has been used. The B cells would not even benefit from the use of an irregular **Generate** statement as each one would be irregular. The indirect method of instantiation has been used.

Remember that the circuit that results from this description can itself be a component. This is therefore an example of a highly hierarchical design with multiple levels of external blocks.

```
entity CLA_CELL_A is
    port (A, B, C : in BIT; S, P, G : out BIT);
end CLA_CELL_A;
architecture DATAFLOW of CLA_CELL_A is
begin
    -- Level one (top level) logic
    S <= A xor B xor C;
    P <= A or B;
    G <= A and B;
end DATAFLOW;

entity CLA_CELL_B is
    port (    G1, G2, P1, P2, C1 : in BIT;
              G3, P3, C2 : out BIT);
end CLA_CELL_B;
architecture DATAFLOW of CLA_CELL_B is
begin
    -- Level two (Intermediate level) logic
    C2 <= G2 or (P2 and C1);
    G3 <= G1 or (P1 and G2);
    P3 <= P1 and P2;
end DATAFLOW;

entity CLA_CELL_C is
    port ( G, P, C1 : in BIT; C2 : out BIT );
end CLA_CELL_C;
architecture DATAFLOW of CLA_CELL_C is
begin
    C2 <= G or (P and C1);
end DATAFLOW;
```

Figure 6.11 Architectural descriptions of the cells in the CLA adder.

6.2.3 Synthesis and optimization

Large blocks of arithmetic hardware usually determine the critical timing path of a circuit. The addition and subtraction circuits shown in this section may therefore be part of this critical path and must be subjected to appropriate timing constraints during their design. These are in addition to any area constraints that are also imposed. The optimization of the examples in this section therefore consists of the following.

- An area optimization stage in which low, medium and finally high levels of effort are applied. The statistics noted in the following tables are for the high level of effort. This always produced the best result.

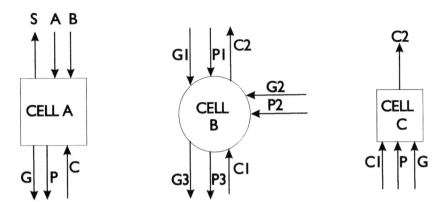

Figure 6.12 'Generic' component I/O structure for the 4-bit adder.

- A performance (speed) optimization stage in which predefined timing figures had to be achieved. For each adder circuit, the critical path (input to output) had to be less than 5ns. To compare the structures of circuits with this performance, the area was unconstrained. A low level of performance optimization effort was always applied first. If the constraint could not be met, medium and then high levels were tried. Only in one case did a circuit fail to meet its target. The level of effort required for each circuit has not been recorded.

As before, the adder types are separated into three sections – Integer, RC and CLA. Within each section both the synthesis and optimization results will be discussed together.

Synthesis and optimization of the integer adder

The simplest adder, INTGR_ADDER-ONEADD, demonstrated a signed integer addition operation. The circuit in Figure 6.14 shows how the simple addition operator is translated into hardware. The key features of this circuit include:

- the integer range –8 to 7 has been transformed into a 5-bit, sign-magnitude binary representation;
- a ripple-carry structure has been created. This can be seen from the way the output bits are progressively generated at each stage of the calculation and from the Nand and Xor gate structure.

The synthesizer will always choose to implement an integer adder in this manner, regardless of the number of bits. As timing constraints for a particular design are not known at the synthesis stage, a circuit with minimum area will be aimed for. Table 6.1 shows the statistics of this circuit. As

```
library ARITH;
use ARITH.ARITH_TYPES.SIGNED;
entity CLA is
   port (    A, B : in SIGNED(3 downto 0);
          OVERFLOW : out BIT;
          S : out SIGNED(3 downto 0));
end CLA;

architecture STRUCTURE of CLA is

   component CLA_CELL_A
      port ( A, B, C : in BIT; S, P, G : out BIT );
   end component;
   component CLA_CELL_B
      port (    G1, G2, P1, P2, C1 : in BIT;
             G3, P3, C2 : out BIT);
   end component;
   component CLA_CELL_C
      port ( G, P, C1 : in BIT; C2 : out BIT );
   end component;
   signal C : BIT_VECTOR(0 to 3);
   signal P, G : BIT_VECTOR(0 to 5);

begin
   C(0) <= '0'; -- force carry in to be '0'
   ACELLS : for K in 0 to 3 generate
      CELL_AK : CLA_CELL_A port map (A => A(K), B => B(K), C => C(K),
                                   S => S(K), P => P(K), G => G(K) );
   end generate;
   CELL_B0 : CLA_CELL_B port map (   C1 => C(0),   C2 => C(2),
                              P1 => P(5),    P2 => P(4), P3 => open,
                              G1 => G(5),   G2 => G(4),G3 => open );
   CELL_B1 : CLA_CELL_B port map (   C1 => C(0),      C2 => C(1),
                              P1 => P(1),    P2 => P(0), P3 => P(4),
                              G1 => G(1),   G2 => G(0),G3 => G(4) );
   CELL_B2 : CLA_CELL_B port map (   C1 => C(2),      C2 => C(3),
                              P1 => P(3),    P2 => P(2), P3 => P(5),
                              G1 => G(3),   G2 => G(2),G3 => G(5) );
   FLOWOVER : CLA_CELL_C port map ( C1 => C(3), P => P(3),
                                 G => G(3), C2 => OVERFLOW );
end STRUCTURE;
```

Figure 6.13 CLA adder structural architecture.

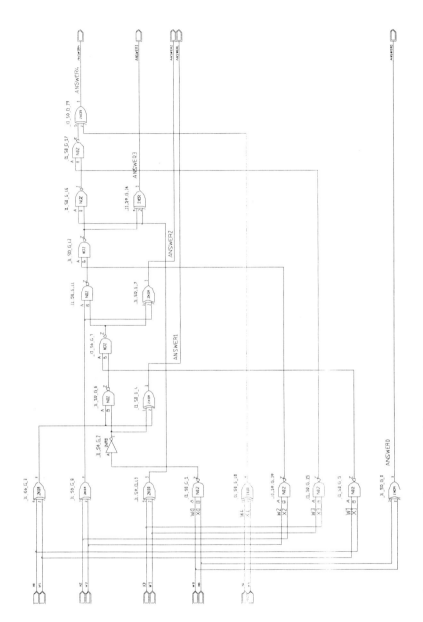

Figure 6.14 Synthesized circuit of the integer adder.

Table 6.1 Logic synthesis and optimization statistics for the integer adder and adder/subtractor architectures. Timing constraints were met in each case. No final area constraint was applied

Parameters Temperature: 25.0 °C Voltage : 5.0 V	Synthesis Library: generic	Optimization Type: area Level: L, M, H	Optimization Type: speed Max. delay 5 ns
Circuit: INTGR_ADDER-ONEADD			
Cells	20	No change	26
Transistors	132	No change	228
Area	16 724	No change	25 702
Longest input to output delay	5.981 ns	No change	4.025
Circuit: INTGR_ADDSUB-DATAFLOW			
Cells	60	45	42
Transistors	364	244	360
Area	45 770	29 398	39 080
Longest input to output delay	8.099 ns	8.141 ns	4.631 ns

Table 6.2 Logic synthesis and optimization statistics for the ripple carry adder architectures and the signed dataflow structure. Timing constraints were met in each case. No final area constraint was applied

Parameters Temperature: 25.0 °C Voltage: 5.0 V	Synthesis Library: generic	Optimization Type: area Level: L, M, H	Optimization Type: speed Max. delay 5 ns
Circuit: ADD-RIPPLE1 ADD-RIPPLE2			
Cells	39	40	41
Transistors	262	200	272
Area	31335	24997	31335
Longest input to output delay	14.835 ns	7.993 ns	4.866 ns
Circuit: ADD-RIPPLE3			
Cells	20	29	29
Transistors	148	134	146
Area	17 780	16 900	17 956
Longest input to output delay	7.0436 ns	6.020 ns	4.498 ns
Circuit: SIGNED_ADDSUB-DATAFLOW			
Cells	40	44	53
Transistors	252	248	436
Area	30 631	29 574	47 882
Longest input to output delay	10.258 ns	9.954 ns	4.963 ns

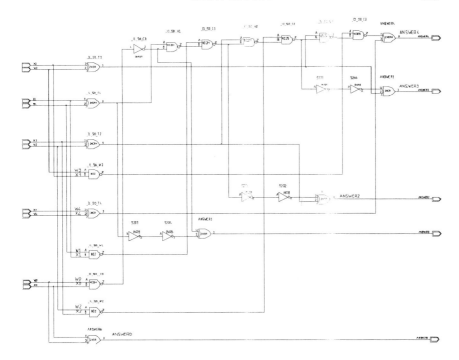

Figure 6.15 Optimized circuit of the integer adder.

the results for comparable circuits show, such as the architecture **RIPPLE3** in Table 6.2, the transistor count and area figure of the **ONEADD** architecture are the best achieved by any adder circuit here. This is despite the additional bit of the sign-magnitude representation. Attempts at area optimization failed to produce any improvement, further indicating that the synthesis process has already performed the best area translation.

Performance optimization required a reduction in critical path delay of only 0.981ns. As characterizes most of the performance optimization processes in this section, this is always accompanied by an increase in area. Figure 6.15 shows the optimized circuit. It is characterized by:

- additional buffers (inverter pairs), which speed up critical nodes throughout the network.
- certain critical gates exchanged for ones with higher drive capability, for example a **2XOR** is replaced with a **2XORS** or **2XORH**.

These modifications increase the area of the circuit without causing a significant restructuring. They are generally the first transformations that are attempted when making improvements to the performance of a circuit. Table 6.1 shows that the delay along the critical path in the optimized circuit is within the specified upper bound at a cost of 96 additional transistors.

The same table shows the data for the INTGR_ADDSUB-DATAFLOW model. This is an integer addition or subtraction circuit, the function of which is selected by the MODE input signal. As the statistics indicate, the synthesized circuit occupies almost three times the area as the single adder circuit. This consists of three elements:

- the addition logic;
- the subtraction logic;
- the selection logic.

The synthesis process allowed resources to be shared between the two operators but this does not seem to have occurred. The sharing of resources between dissimilar operators such as + and – is discussed in the Dos and Don'ts section. Area optimization has enabled a significant reduction in logic to occur. This of course involved the removal of redundant and duplicated gates. The increase in transistor count, back to the level of the original synthesized circuit, is not due to the return of any of this logic. Instead, the performance-enhancing measures, as discussed above, that have reduced the critical path length by half, are the cause. The synthesized and optimized circuit diagrams are not shown.

Synthesis and optimization of the ripple-carry adder

The ripple-carry functions contained in the package ARITH_TYPES have been used in four architectures. RIPPLE1 and RIPPLE2 call the same RIPPLE function but the latter does this indirectly through an overloaded '+' operator. Although a ripple-carry structure was used to implement the integer addition operator, the statistics in Table 6.2 indicate that the user-defined versions are double the size and less than half the speed. RIPPLE1 and RIPPLE2 also have an overflow output of course, which can account for some of this additional logic. However, the synthesizer cannot implement user-defined routines as efficiently as predefined ones, as it does not 'know' exactly what they are. However, the optimizer can achieve area and then significant performance gains for these two circuits, as Table 6.2 shows.

Figure 6.16 shows the synthesized circuit, which appears quite dissimilar to the integer ripple-carry circuit of Figure 6.14. In particular, there are many more levels of logic. These are a result of the implementation of the carry equation using two-input Or and And gates as specified in the function. By unrolling the function's loop, four copies of the necessary hardware required to implement the sum and carry equations have been created. Figure 6.17 shows the circuit after area and performance optimization. Unlike the integer adder, this circuit has been heavily restructured. Again, higher drive logic gates have been used in certain places, along with occasional buffers. However, the major changes are the replacement of slow logic structures with faster ones. For example an Xor gate can be replaced with And and Or gates with inverters to produce the same func-

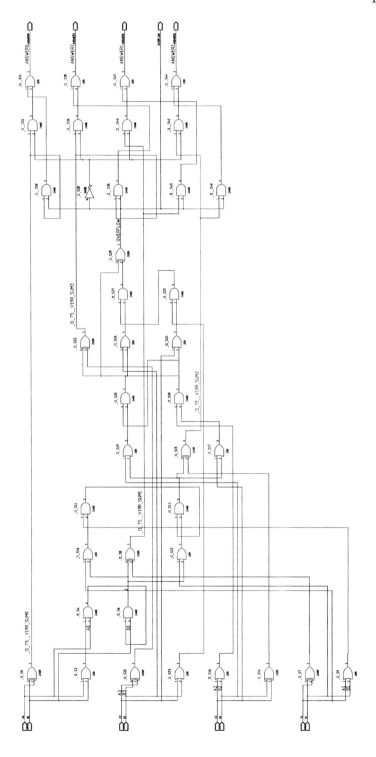

Figure 6.16 Synthesized circuit for the RIPPLE1 and RIPPLE2 architectures.

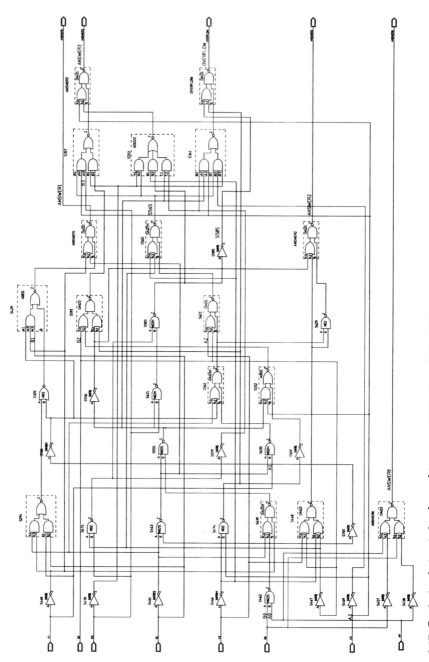

Figure 6.17 Optimized circuit for the architectures RIPPLE1 and RIPPLE2.

tion. The optimizer has determined that, in this case, the resulting circuit will be faster although larger.

RIPPLE3 is a simpler architecture, although the function it calls is more flexible than the function called in RIPPLE1, for example. In this latest architecture, the ripple-carry function can add or subtract the 2's complemented binary numbers. However, in this case it is called with a constant, '0'. This means that the function will only ever be used for addition and so only the required elements of the function need to be synthesized. The statistics in Table 6.2 show that the circuit produced is almost as small as the integer adder circuit.

One of the benefits of using a function over a component is this 'synthesize as necessary' approach, which was also discussed earlier. It is often the case that functions, particularly those provided by tool vendors, do not need to be synthesized at all. These **built-in** functions are useful for bridging the gap between hardware 'reality' and software 'representation'. Box 6.10 explains further.

The synthesized and optimized circuits for the RIPPLE3 architecture are not shown. However, as the statistics indicate, the synthesized circuit is similar to but smaller than the RIPPLE1 and RIPPLE2 circuits. Also, the optimization of the circuit followed a similar pattern, with area falling and then rising again as the optimizer aimed to meet its timing constraints. Similar transforms occurred in the circuit with a reduction of levels and the exchange of Xor gates for And/Or gate equivalents.

Box 6.10 How to use built-in functions

Built-in functions are not compiled into hardware by the synthesizer. They allow the designer to overcome problems at the abstract language level, such as type mismatches between objects, which do not occur at the gate level. Such problems would cause errors when a design is simulated but not synthesized. A function that is declared as built in, by using the BUILT_IN attribute (Chapter 7), can be called as many times as required in a design without incurring any hardware costs.

Most vendor-supplied built-in functions are designed to be flexible by specifying array parameters as unconstrained. The user can therefore call these functions using suitable type array objects of any length. Inside the function, attributes can then be used to determine the RANGE or LENGTH of an array. A user-defined function may also employ this technique to allow greater flexibility from the function. However, as these functions are more likely to be synthesized, multiple calls using different length parameters are likely to lead to a significant hardware overhead.

In the final architecture that uses the user-defined ripple-carry function both addition and subtraction are possible. SIGNED_ADDSUB contains a call to the same RIPPLE function used in RIPPLE3. This time, however, the operation is selected externally by the MODE signal and therefore the full function must be synthesized. This architecture is included so that the circuit can be compared with the integer addition and subtraction architecture, INTGR_ADDSUB.

The primary difference between the two implementations is that the integer version creates a circuit using sign-magnitude binary representation whereas the ripple-carry version uses 2's complement. As stated earlier, the former must contain logic for the addition, subtraction and control/ select. The latter, however, uses the same logic for addition and subtraction, except that the second one requires the bits of B to be inverted and the carry in to be set. The arithmetic logic is therefore reduced and the control logic a little more complex.

Comparing the synthesis statistics for these two circuits from Tables 6.1 and 6.2 immediately reveals the saving in area that is achieved with the user-defined arithmetic function. It is noticeable from the start, however, that the additional control logic of the user-defined version results in a slower circuit – area has been traded off for speed. The synthesized circuit is shown in Figure 6.18. It is surprisingly compact for a 4-bit addition *and* subtraction circuit.

Unfortunately, this circuit does not optimize well. Area optimization makes only a slight impression on the circuit. The performance optimization process then reduces the number of logic levels to meet the timing constraint but in doing so causes a significant increase in area. The final delay and area statistics of the circuit are worse than those for the integer adder and subtractor. Why this should occur is not entirely clear. As the optimized circuit diagram in Figure 6.19 shows, a wide range of transformations have been made in the circuit but the levels of logic have not decreased significantly, forcing the optimizer to resort to more high drive components. These are faster but consequently much larger.

Synthesis and optimization of the carry-lookahead adder

The carry-lookahead adder is built from three components and is itself designed to be a component within another design. It therefore has a strong hierarchy that the designer can make use of during optimization. By optimizing the area or performance of one component, each instance of that component is also optimized. This could mean a vast saving in computing effort for a component with several instances. On top of that, a small improvement in one component may mean a significant improvement in the circuit as a whole.

These benefits are demonstrated here. The top level of the carry-lookahead circuit, defined by the entity CLA, is dominated by intercon-

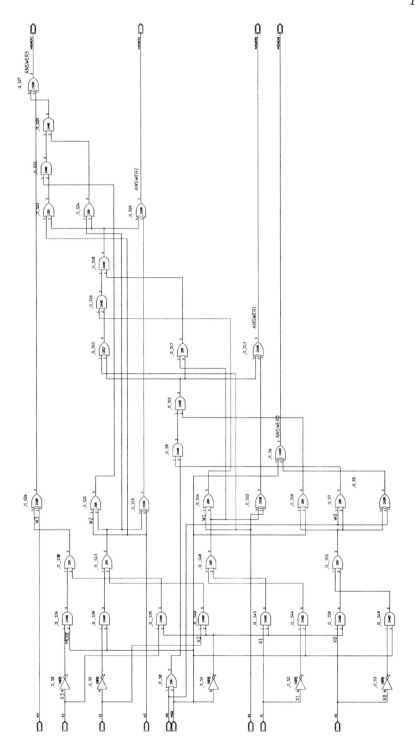

Figure 6.18 Synthesized circuit of the signed binary addition/subtraction description SIGNED_ADDSUB.

194

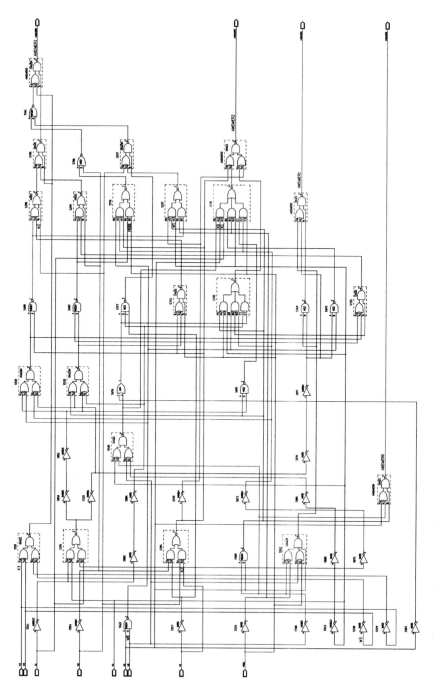

Figure 6.19 Optimized circuit for the signed binary addition/subtraction description SIGNED_ADDSUB.

nect. This links four instances of cell A with three of cell B and one of cell C, as shown in Figure 6.20. Although the layout of the blocks is different to the original algorithmic diagram that was shown in Figure 6.10, the pattern of interconnect is identical and the instance name of each block enables the structure to be verified. In each component block a realistic critical path delay of 0.8ns was specified as an optimization timing constraint. As each synthesized component was very small it was possible to see that redundant or duplicated logic did not exist. Therefore area optimization was not attempted before performance optimization. A summary of their synthesis and optimization statistics are shown in Table 6.3.

The synthesized versions of these circuits have been omitted as their structure can be determined easily from the Boolean equations in their architectures. For example, cell A consists of two Xor gates, an And and an Or gate. Figure 6.21 shows the final optimized circuit for this cell, which has a failing critical path. As the circuit shows, the critical path is from A or B to S and the problem therefore lies with the speed of the Xor gates. This path was restructured by the optimizer using a peephole transformation. This breaks apart the large, slow gates and replaces them with more – smaller and faster – equivalent structures. This same transform occurred in the optimized circuits of Figures 6.17 and 6.19 but this is a clearer example. After the restructuring the optimizer still found that it was unable to

Table 6.3 Logic synthesis and performance optimization statistics for the carry-lookahead arithmetic architecture's components. Where timing constraints were not met (despite low, medium and high levels of effort) the statistics are shown in bold

Parameters Temperature: 25.0 °C Voltage: 5.0 V	Synthesis Library: generic	Optimization Type: speed Max. delay 0.8 ns
Circuit: CLA_CELL_A-DATAFLOW		
Cells	4	10
Transistors	32	71
Area	3872	13 555
Longest input to output delay	1.589 ns	**0.851 ns**
Circuit: CLA_CELL_B-DATAFLOW		
Cells	5	8
Transistors	30	34
Area	3520	4401
Longest input to output delay	1.044 ns	0.657 ns
Circuit: CLA_CELL_C-DATAFLOW		
Cells	2	2
Transistors	12	14
Area	1408	1584
Longest input to output delay	1.044 ns	0.594 ns

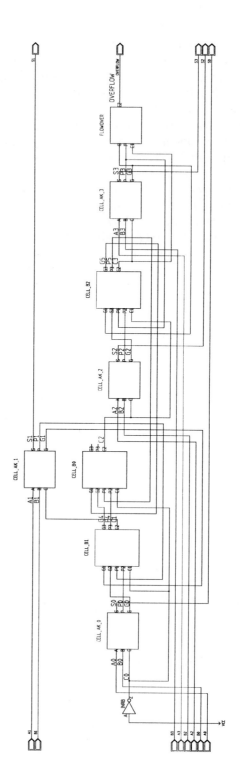

Figure 6.20 Top-level schematic of the carry-lookahead adder.

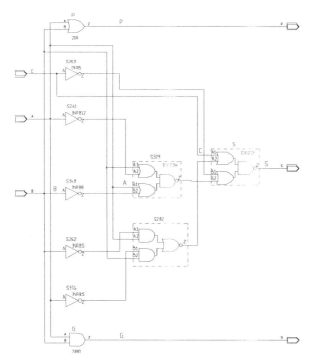

Figure 6.21 Optimized circuit for cell A.

meet the constraints and so it replaced several of the most critical gates with faster versions. The inverter on input C is not one of these, for example, and has not been enhanced.

Cells B and C have achieved the required timing constraint but the former, in particular, has undergone significant modifications. Two main transformations have occurred as follows:

- The paths through to G3 and C2 have had peephole restructuring.
- A DeMorgan transformation has be used on P3. Also known as complementation, this involves the replacement of the And gate with a Nor gate and input inverters. Again, these new gates are smaller and faster.

The optimized circuits for cells B and C are shown in Figures 6.22 and 6.23.

Table 6.3 indicates that in each of the three cells the area of the circuit has increased with performance optimization. In cell A this has been an over fourfold increase, which would probably be unreasonable in a real design situation, and suggests that a less stringent timing constraint should be aimed for or a redesign attempted.

The statistics for the whole carry-lookahead adder are shown in Table 6.4. This is a special example in that there are two synthesized versions of

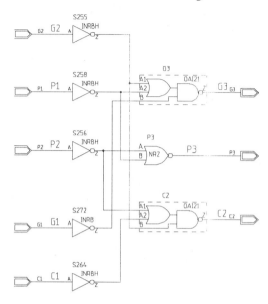

Figure 6.22 Optimized circuit for cell B.

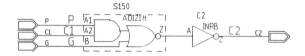

Figure 6.23 Optimized circuit for cell C.

the circuit. The first uses the synthesized components and the second uses the optimized components. The latter would normally be used, just as in the previous structural examples, but both are shown to enable comparisons to be made.

The best circuit to compare the carry-lookahead adder with is the ripple-carry architecture, RIPPLE I, shown in Table 6.2, as both have 4-bit operation and output an overflow signal.

When using only synthesized components the carry-lookahead adder produces a smaller and faster circuit than the ripple-carry adder. Using optimized components while retaining the top-level structure should bring the critical path delay down to a satisfactory level. It does not reach the 5ns mark, however, which is very disappointing when the cost in area is so high – this is the largest optimized or unoptimized arithmetic circuit by far. As we have maintained a modular structure to the design it is possible to examine closely not just the gates but also the components on the critical path. Figure 6.10 showed the original structure of the algorithm. From this it is relatively easy to determine the critical path. It will of course start with the least significant input bit and finish with the most significant output bit – A_0 or B_0 to S. Therefore the critical path should go through cells A0, B1, B0, B2 and A3. In fact, the statistics that the synthesis tool produces confirm this, as shown in Table 6.5. This table shows the components, gates and pins on the gates that are responsible for the critical delay in the circuit. Using this information the designer can address precisely the problem areas of the circuit at a component level. For example the long delays through the gates C2 and A2 in cells B0 and B2 need considering. The advantage of the component approach is that looking at a problem in one cell can immediately remedy a similar problem in another.

The third column of statistics in Table 6.4 is for the flattened, performance optimized version of the synthesized circuit that uses the optimized components. The critical path delay has been reduced to within the de-

Table 6.4 Logic synthesis and performance optimization statistics for the carry-lookahead arithmetic architecture

Parameters Temperature: 25.0 °C Voltage: 5.0 V	Synthesis Library: generic Components: synthesized	Synthesis Library: generic Components: optimized only	Optimization Type: speed Max. delay 5 ns
Circuit: CLA-STRUCTURE			
Cells	34	67	29
Transistors	230	652	150
Area	27 638	69 183	18 308
Longest input to output delay	5.965 ns	7.116 ns	4.657 ns

Table 6.5 Delays and gates along the critical path of the carry-lookahead adder. Bold figures are discussed in the text. R/F means the delay on a rising/falling input signal edge

Component	Gate	I/O pin	Type	Maximum delay (ns)	Capacitance (pF)
CELL_AK_0	P	B-Z	OR2	0.538 R	0.369
CELL_B1	S256	A-Z	INRBH	0.389 F	0.464
CELL_B1	P3	A-Z	NR2	0.695 R	0.378
CELL_B0	S256	A-Z	INRBH	0.515 F	0.464
CELL_B0	C2	A1-Z	OAI21	**1.577 R**	0.910
CELL_B2	S264	A-Z	INRBH	0.535 F	0.262
CELL_B2	C2	A2-Z	OAI21	**1.552 R**	0.910
CELL_AK_3	S263	A-Z	INRB	0.979 F	0.371
CELL_AK_3	S	B1-Z	OAI22H	0.336 R	0.298
				7.116 R	

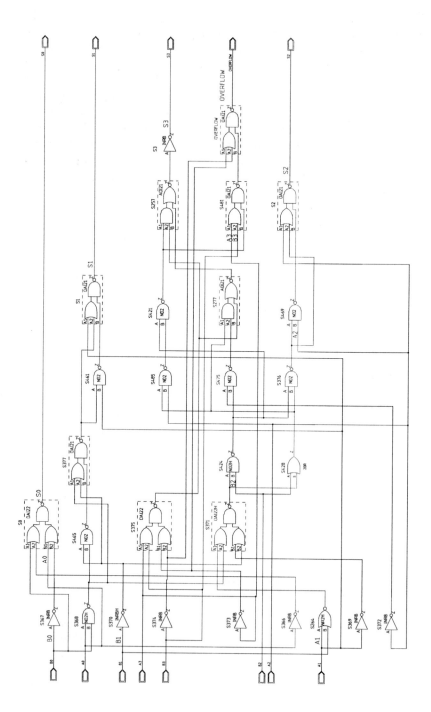

Figure 6.24 Optimized (flattened) carry-lookahead circuit.

sired limit and the area has been reduced over fourfold. Although the hierarchical structure has gone, the carry-lookahead algorithm is still being
implemented. The strange characteristic of this circuit is not that it is no-
ticeably faster than any of the other similar adders, but that it achieves
comparable performance with a much smaller area. The optimized circuit
is shown in Figure 6.24. The massive reduction in area of this circuit suggests a large amount of redundancy and duplication in the components.
This situation will occur in any circuit that employs a large number of small
components.

6.2.4 Adder summary

The primary language elements introduced through these arithmetic circuits have been functions, loops and overloading. The characteristics of
functions are different to those of components and each is flexible in its
own way. Functions benefit from their portability but are limited in their
N-in 1-out structure.

Overloading is a technique that enables the designer to choose the version of a function or operator that suits the object parameter types and
formats. It is a convenient way of allowing tool vendors or users to extend
the capabilities of the language. This will, of course, place restrictions on
the portability of a design.

The arithmetic circuits in this section were primarily examples of three
types of adder. Their optimized circuits demonstrated that, although it is
often possible to obtain reasonable performance, the cost in area terms may
be significantly higher for user-defined circuits than for those that are built
in. In particular, the circuit produced by the synthesis of a built-in operator, such as the integer adder, is area efficient and fast. The carry-lookahead
adder was mainly used as a demonstration of how a rigid algorithm can be
mapped into a few components and the savings in time and effort that this
approach brought during synthesis and optimization.

6.3 2 ×2 PARALLEL MULTIPLIER

This section presents the design of a 2 ×2 parallel multiplier that employs
the Booth recoding algorithm. The multiplier very often accounts for the
largest block of logic in the circuit and therefore only a small example can
be presented. During the initial investigations when examples were being
selected for this text, an 8 ×8 parallel multiplier was synthesized. The pr ob-
lem with using this circuit would have been how to illustrate clearly the
function of the circuit when it was spread over seven pages of schematic!

The Booth multiplier was selected as it is relatively simple and allows
the previously defined ripple-carry adder/subtractor to be used. Also, it is
a good method of designing multiplier hardware that must operate on
signed numbers.

6.3.1 Functional overview

A multiplier based on the Booth recoding scheme is shown in Figure 6.25. The operation of this multiplier can be summarized as follows:

1 Load A and B into the registers and clear P. Assume AX(0) is '0'.
2 Examine AX(1) and AX(0).

  If AX(1)='0' and AX(1)='0' then add '0' to P.
  If AX(1)='0' and AX(1)='1' then add B to P.
  If AX(1)='1' and AX(1)='0' then subtract B from P.
  If AX(1)='1' and AX(1)='1' then add '0' to P.

3 Arithmetically shift the contents of P and AX right by one bit (preserving the sign of P).
4 Repeat from step 2 for all n bits of A.

  The n+1 bit result can then be extracted from bit 0 of P and bits n downto I of AX. In this case n=2, so the result will be 3 bits long including the sign.
  To implement the above operations the circuit must contain mechanisms to:

• add and subtract B from P; and
• shift P and AX arithmetically right.

  The multiplier is not pipelined and therefore a complete parallel multiply operation must be complete within one machine cycle.

6.3.2 VHDL code design

The multiplier could be implemented as a function or as part of a process that will form a component. A function would probably be the best solu-

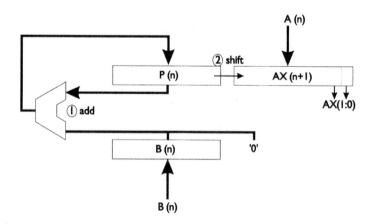

Figure 6.25 Structure of a simple Booth parallel multiplier.

tion and would take a similar form to the RIPPLE functions in the ARITH_TYPES package. Using a function would enable a generic multiplier to be created and would allow the dataflow element – the actual calculation – to be separated from the register transfer element. This occurred with the ripple-carry architectures and functions. This approach is left to the reader to implement.

By embedding the multiplication process within an architecture and a process, the sequential (storage) elements are mixed with the dataflow elements of the design. The entity and architecture for the parallel multiplier, PMULT-BOOTH is shown in Figure 6.26. The Library and Use clauses provide access to the SIGNED type declaration, the ripple-carry adder/subtractor and a package called SIGNED_UTILS that will be discussed in more detail later.

The whole multiply routine is inside one process, MULT_OP, which is controlled by a Wait statement and the external signal, PHI. This could be a clock signal but may equally be any signal that tells the multiplier that data is present and that the multiplication process may begin. Any signal assignments within the process will infer sequential logic elements. To avoid this occurring as much as possible, a number of variables are declared (and defined before being used) in the process to store the intermediate calculation values around the loop. As a result, the only signal assignment that occurs is to the output port RESULT at the end of the loop.

So let us look at the multiplication operation. First, another type of loop is being used. The Infinite Loop is not always supported by synthesizers owing to the unpredictable nature of their operation – the loop could be unrolling forever if a condition that causes an exit is never met. Similarly, the While Loop is often not supported for synthesis. The syntax of both these alternative loop types is shown in Box 6.11. This loop is controlled by the counter I, which increments on each pass around the loop. When I reaches the n – the number of bits in A – the Exit statement quits the loop. This statement does not necessarily require the loop label in this example but one should be used when embedding loops to clarify which loop is to be stopped. Next is another loop control statement that will immediately start the next iteration of the loop, assuming its condition is met. They are both discussed in Box 6.12.

Inside the loop MODE is set depending on the state of the two leftmost bits of AX. This is then used first to determine whether an addition or subtraction operation should occur and, if so, to indicate to the function RIPPLE which should be performed. As the operation of this function depends on the value of MODE(0), the control logic within RIPPLE that selects add or subtract must be fully synthesized.

The new value of P is then concatenated with AX and the result, S, is shifted arithmetically right by 1 bit. The '&' operator is not defined for the SIGNED type of object with which it is being used. Therefore the package SIGNED_UTILS, shown in Figure 6.27, is accessed by the multiplier. This

```vhdl
library ARITH;
use ARITH.ARITH_TYPES.all, ARITH.SIGNED_UTILS.all;
entity PMULT is
   port (   A, B : in SIGNED(I downto 0);
            PHI : in BIT;
            RESULT : out SIGNED(2 downto 0) );
end PMULT;
architecture BOOTH of PMULT is
begin
   MULT_OP : process
      variable P : SIGNED(I downto 0);
      variable S : SIGNED(4 downto 0);
      variable AX : SIGNED(2 downto 0);
      variable MODE : BIT_VECTOR(I downto 0);
      variable I : INTEGER range A'reverse_range; -- loop counter
   begin
      wait until (PHI='0');            -- wait for 'go' signal
      P := (others=>'0');
      AX := A & '0';
      I := 0;
      CALCULATE : loop                 -- loop 'forever'
         case AX(I downto 0) is        -- examine low bits
            when "01" =>    MODE := "10";   -- add
            when "10" =>    MODE := "11";   -- sub
            when others => MODE := "00";    -- null
         end case;
         if MODE(I) = "I" then
            P := RIPPLE(P, B, MODE(0));       -- addition/subtraction
         else
            P := P;           -- 'add' zero
         end if;
         S := P & AX;              -- concatenate
         S := ARITH_SHIFT_RIGHT(S, I);    -- shift
         P := S(4 downto 3);            -- reassign to P
         AX := S(AX'range);            -- reassign to AX
         I := I + I;                   -- inc. counter
         exit CALCULATE when I=A'length;       -- exit or loop
      end loop;
      RESULT <= S(3 downto I);
   end MULT_OP;
end BOOTH;
```

Figure 6.26 VHDL code for the 2 ×2 parallel multiplier .

Box 6.11 How to create While and Infinite Loops

While Loop statement

1 A While Loop statement enables a set of other sequential statements to be repeated until a specified condition is met. It is not always supported for synthesis. It has the syntax:

```
while_loop_statement ::=
   [ loop_label : ]while condition loop
           sequential_statements
   end loop [ loop_label ];
```

2 Any sequential statements can be used inside the loop *except* a Wait statement. Therefore a While Loop can only be used to create combinational logic. This is constructed by unfolding the loop during compilation.

3 A nested loop must be fully contained within its outer loop. It is advisable to use loop labels for identification when nesting loops.

4 By acting on a condition inside the loop the present iteration of the loop or the whole While Loop statement can be terminated. See Box 6.12.

Infinite Loop statement

1 An Infinite Loop statement enables a set of other sequential statements to be repeated forever. Because of its unpredictable nature it is not often supported for synthesis. It has the syntax:

```
infinite_loop_statement ::=
   [ loop_label : ] loop
           sequential_statements
   end loop [ loop_label ];
```

2 Any sequential statements can be used inside the loop *except* a Wait statement. Therefore an Infinite Loop can only be used to create combinational logic. This is constructed by unfolding the loop during compilation.

3 A nested loop must be fully contained within its outer loop. Use loop labels for identification when nesting loops.

4 An infinite loop must contain a termination mechanism (Box 6.12).

Box 6.12 How to control Loops

The execution of For, While and Infinite loops can be controlled with two particular sequential statements – **Next** and **Exit**.

The Next statement

The **Next** statement forces the remaining sequential statements in a loop to be skipped and the next iteration of a loop to begin. It has the syntax:

next_statement ::=
    [label:] **next** [loop_label] [**when** condition];

1 The optional label is the label associated with the next statement.
2 The optional loop label refers to the loop that must begin its next iteration. The loop label must only be used inside the actual loop to which it refers. The example below illustrates how one **Next** statement can cause one loop to skip to the next iteration and others to restart altogether.
3 Without a loop label the loop at the present nested level begins its next iteration.
4 See the Dos and Don'ts section.

```
L1 : for N in 1 to 4 loop
        L2 : while S/="10011" loop
            next L1 when S="11101";-- next iteration of L1, restart L2
            ...
            next when S="00001";-- same iteration of L1, next iteration of L2
            ...
        end loop L2;
        next when S="01100";                -- next iteration of L1, L2 not active
        ...
    end loop L1;
```

The Exit statement

The **Exit** statement forces the remaining iterations of a loop to be skipped completely. It has the syntax:

exit_statement ::=
    [label:] **exit** [loop_label] [**when** condition];

1 The **Exit** statement operates in a similar manner to the **Next** statement except that the loop is terminated rather than starting its next iteration. Any loops nested inside the terminated loop are also terminated.
2 An **Exit** statement must be used inside an Infinite Loop statement.

```
library ARITH;
use ARITH.ARITH_TYPES.SIGNED;
package SIGNED_UTILS is
  function "&"(A, B : SIGNED) return SIGNED;
  function "&"(A : SIGNED; B : BIT) return SIGNED;
  function ARITH_SHIFT_RIGHT(A :SIGNED; STEP :INTEGER) return
SIGNED;
end SIGNED_UTILS;
package body SIGNED_UTILS is
  function "&"(A, B : SIGNED) return SIGNED is
    variable C : SIGNED(A'length+B'length-1 downto 0);
  begin
    C(C'left downto B'length+1) := A;
    C(B'range) := B;
    return C;
  end "&";
  function "&"(A : SIGNED; B : BIT) return SIGNED is
    variable SIGNED_B : SIGNED(0 downto 0);
  begin
    SIGNED_B(0) := B;
    return ("&"(A,SIGNED_B));
  end;
  function ARITH_SHIFT_RIGHT(A:SIGNED; STEP:INTEGER) return
SIGNED is
    variable VA : SIGNED(A'range);
  begin
    VA := A;
    for K in 1 to STEP
    loop
            VA(A'left-1 downto 0) := VA(A'left downto 1);
    end loop;
    return VA;
  end ARITH_SHIFT_RIGHT;
end SIGNED_UTILS;
```

Figure 6.27 Functions for the type SIGNED used in the multiplier example.

contains the overloaded version of the '&' operator and the shift function
ARITH_SHIFT_RIGHT. In fact, two versions of the overloaded operator exist. One takes two SIGNED type parameters and the other, one SIGNED parameter and a BIT (or bit literal in the multiplier example). The shift function is a very simple realignment operation but has been created as a more general routine that can shift right any number of bits. This is achieved with a loop. Remember that signal assignments cannot be made inside a

function, so in every case incoming signals must be assigned to variables. These are manipulated in the function and one is returned as the result. Back in the multiplier loop, after the shift operation, the intermediate value S is reassigned to the 'registers' P and AX, ready for the next iteration or the exit condition to be TRUE.

6.3.3 Synthesis and optimization

Synthesis results

The synthesis of the multiplier will involve the unrolling of the loop. A 2 ×2 multiplication means that two identical blocks of har dware will be created to represent the operations in the loop. Just as the synthesizer did not 'know' that it was creating adder circuits with the user-defined ripple-carry adder examples, it does not 'know' that this circuit is a multiplier. Instead, it simply carries out a series of addition and shift operations. Area optimization is therefore the only way to spot common elements and combine them.

The ripple-carry arithmetic function called in the multiplier loop was used in the same manner in the design example SIGNED_ADDSUB-DATAFLOW of Figure 6.9. In that case 4-bit addition/subtraction occurred whereas here it is only 2 bit. However, two instances of this arithmetic function will occur in the circuit because of the loop. It can therefore be estimated from the statistics in Table 6.2 and the new statistics in Table 6.6 that over half the area of the synthesized multiplier is due to the addition/ subtraction block. The table also shows the input to latch delay of the circuit. The synthesized circuit, illustrated in Figure 6.28, shows that the only latches occur at the output of the circuit. This is the intended situation, as discussed during the design process, allowing the registered section to be separated from the dataflow section. Of course there would be input registers elsewhere that store the values of A and B. Given the timing statistic in the table, the set-up time for this input data in the synthesized circuit is around 21ns.

The latches for RESULT(1) and RESULT(2) have multiplexed inputs. The occurrence of these could not have been predicted from the code. Their data select signal is high when MODE(1) is high and therefore the synthesizer has decided to use these latches to select bits for RESULT from either the addition/subtraction operation or the old value of P.

This circuit illustrates the contents of the MULT_OP process only. As there is no other logic outside of this, the top-level schematic block has not been shown.

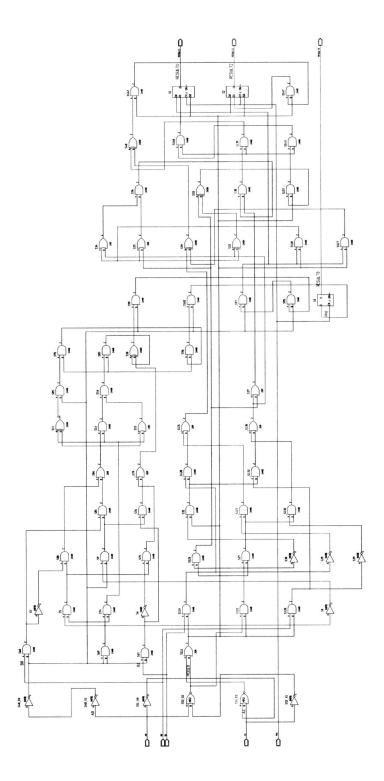

Figure 6.28 Synthesized circuit for the 2 ×2 multiplier.

Table 6.6 Logic synthesis and area optimization statistics for the Booth 2 ×2 multiplier architecture. Resources were not shared and performance optimization did not reduce the delay

Parameters Temperature: 25.0 °C Voltage: 5.0 V	Synthesis Library: generic	Optimization Type: area Max. area: 20 000
Circuit: BOOTH_PMULT-PARALLEL		
Cells	70	16
Transistors	460	146
Area	55 452	17 251
Longest input to latch delay	20.721 ns	2.949 ns

Optimization results

Optimizing this circuit for area produced a very striking result. Low and medium levels of effort reduced the area and delay by over half. A high level of effort produced the results shown in Table 6.6 and the circuit in Figure 6.29. An area constraint of 20000 μm^2 was placed on the circuit, which was easily achieved. The most amazing statistic however is the reduction of the input to latch delay from almost 21ns to less than 3ns. Fr om the circuit diagram the extent of the change can be seen, but why has it occurred?

The VHDL description of the multiplier simply modelled its behaviour – a behavioural architecture. The benefit of this is that the same description can easily be expanded to model larger multipliers. The only changes required are to the lengths of certain vectors. The alternative description would be a dataflow one that used the simple Boolean equation representation for each output. However, with 2^{2n} combinations of inputs, the process of determining these equations is very error prone and time consuming for anything but the smallest multiplier. Unfortunately, unless a particular form of multiplier is required, this second method tends to produce the best circuits.

This is where the optimizer is very useful. If Figure 6.29 is examined, it can be seen that the logic that now remains implements the simple Boolean equations of a 2 ×2 multiplier. For example, RESULT0 is just (A0 and B0), although it is implemented with the faster but equivalent Nand logic. This example has really illustrated the power of the logic optimizer.

6.3.4 Multiplier summary

The Booth recoding algorithm of this multiplier required a behavioural implementation as opposed to the structured approach taken with the carry-lookahead adder. A behavioural style allows the multiplier to be implemented in different ways, such as a function or a process. The former ap-

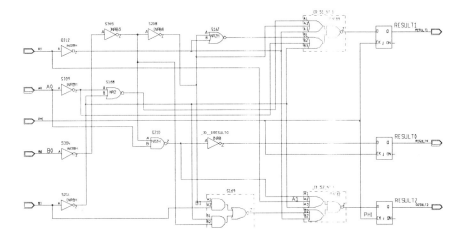

Figure 6.29 Optimized circuit for the 2 ×2 parallel multiplier .

proach has been used extensively in the earlier examples in this chapter and is left for the reader to attempt.

The description of the multiplier is compact because of the **Loop** statement. Of course the designer must always be aware of the amount of hardware that can be produced when this is synthesized. The same applies to arithmetic operators.

With this circuit the optimizer demonstrated how it could convert a designer-oriented or 'user-friendly' description into a hardware-efficient circuit. This is a good demonstration of the optimizer's power when dealing with small circuits. A 16 ×16-bit multiplier would not just take significantly longer to process, but producing a minimized Boolean circuit would be much harder.

6.4 DOS AND DON'TS

6.4.1 Implicit resource sharing

All the relational operators and the arithmetic operators (*, +, −) can share the resources allocated to other instances of the same and sometimes different operators. The VHDL synthesizer will automatically share resources when the option is set at compile time. If this is overridden then all operators will create their own block of logic.

When resources will be shared

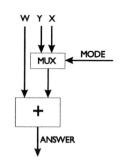

The following are examples of situations when the resources can be automatically shared. For simplicity, all objects are of type INTEGER with a constrained range of –8 to 7.

```
with MODE select
   ANSWER <= W+X when 0,
             W+Y when I;
```

```
with MODE select
   ANSWER <= W+X when 0,
             Y+W when I;
```

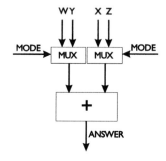

Both these combine a multiplexer with an adder as shown opposite.

```
with MODE select
   ANSWER <= W+X when 0,
             Y+Z when I;
```

This will combine two simple multiplexers with one adder circuit as shown opposite.

```
with MODE select
   ANSWER <= W+X when 0,
             W–Y when I;
```

This will produce a combined adder/subtractor circuit. Not all synthesizers will support this merging of types.

When resources will not be shared

Resources will not be shared automatically in certain situations. If operations are not mutually exclusive, i.e. they may occur at the same time, they will not share resources. There are two examples below.

Sequential :

```
if MODE=0 then
     ANSWER I <= W+X;
else
     ANSWER I <= 3;
end if;
ANSWER2 <= W+Y;
```

Concurrent :

```
ANSWERI <= W + X;
ANSWER2 <= W + Y;
```

Identical operators will always be shared first. The subtractor in the example below will not necessarily be merged with the adders.

```
with MODE select
      ANSWER <= W+X when 0,
                W-Y when I
                W+D when 2;
```

6.4.2 Explicit resource sharing

The examples in section 6.4.1 performed resource sharing automatically if it is allowed. However, to save time and effort on behalf of the synthesis and optimization tools the designer can explicitly share the resource through the use of intermediate terms. The statement

```
if D='0' then
    X <= A * B;
else
    X <= A * 3;
end if;
```

can be replaced with the following, which will explicitly create a single multiplier.

```
if D='0' then
      TEMP := B;
else
      TEMP := 3;
end if;
X <= A * TEMP;
```

6.4.3 Directing the structure with parenthesis

The designer can control the structure of an arithmetic circuit by using parenthesis. Consider the following statement.

```
X <= A + B +C +D;
```

This will create three adders as the addition operations are not mutually exclusive. However, the designer will not know the order in which the operations are being performed – rippling forward, backward and so on. If parentheses are used the designer can direct how the synthesizer tackles this circuit. For example:

X <= A + ((B + C) + D)

This will add B and C, then that result is added to D, and finally that result is added to A. This structure would be useful if A was the last signal to arrive.

6.4.4 Limitations of integer arithmetic

It is difficult to test for overflow in integer arithmetic. Unlike the user-defined functions created in this chapter the integer adder will simply discard any carry out or overflow bit. During simulation a warning is issued if overflow occurs but this is not synthesized into hardware.

Extending the integer range of the result is not a good solution. Look at the example below.

```
entity INTGR_ADDER_OVER is
   port (   W, X : in INTEGER range -8 to 7;
               ANSWER : out INTEGER range -16 to 15 );
end INTGR_ADDER_OVER;

architecture ONEADD1 of INTGR_ADDER_OVER is
begin
   ANSWER <= W + X;
end ONEADD1;
```

The signal ANSWER has been given an extended range so that the output contains an extra bit. However, this fifth bit is not interpreted as a carry bit by the synthesizer and is instead merged with the fourth bit, which it considers is the most significant.

A solution to this problem is to use a manual overflow detection routine as illustrated below.

```
entity INTGR_ADDER_OVER is
   port (   W, X : in INTEGER range -8 to 7;
               ANSWER : out INTEGER range -8 to 7;
               OVERFLOW : out BIT );
end INTGR_ADDER_OVER;

architecture ONEADD1 of INTGR_ADDER_OVER is
begin
   process (W, X)
   variable SUMTEMP : INTEGER range -16 to 15;
   begin
      SUMTEMP := W+X
      if (SUMTEMP < -8) or (SUMTEMP >= 8) then
```

```
        OVERFLOW <= '1';
    else
        OVERFLOW <= '0';
    end if;
    end process;
    ANSWER <= W+X;
end ONEADD1;
```

This is a very area-costly solution and hence it is better all-round to avoid integer arithmetic if features such as overflow detection are required.

6.4.5 The signed and unsigned binary representation of integers

Remember that integers are translated into a binary representation by the synthesizer but the designer must still treat them as integers – bitwise operations on integers are not allowed. An integer with a non-negative range is treated as an unsigned bit vector. An integer that may take a negative value is translated into a bit vector with a sign-magnitude format; it is not 2's complemented.

Also note that a subtype with a non-negative range is still considered to be signed if the type from which it is derived itself has a negative range.

6.4.6 Trapping errors in subprograms

Subprograms (functions and procedures) are stored in packages. A package is compiled before it is used but not synthesized into gates. This means that subprograms can be constructed as modules in which the array lengths and even their precise operation is not known until called. A subprogram will therefore usually compile correctly, assuming there are no syntax errors, but will mask certain errors until the calling architecture body is synthesized. An example of this is could be an attempt to read or assign an array element that does not exist – an out of range error. Until the length of the array is supplied in the parameter list this sort of error will not be trapped.

6.5 EXERCISES

1 Develop a circuit that takes two 4-bit numbers as an input. A 1-bit control signal ADD/SUB determines whether the 5-bit output C is equal to A+B (ADD=1) or A–B (ADD=0) .

2 Design a circuit that takes a 3-bit input A and produces the 9-bit output B, which is equal to A2. Develop behavioural, dataflow and structural designs for this circuit. You should probably use the **Generate** function for the structural description. This could be implemented using a multiply function, but it is possible to create a substantially simpler circuit.

3 In signal processing systems a very common computation is the multi-ply accumulate A= A + B*C. Design such a multiply accumulator, as-suming a 16-bit input A and 3-bit inputs B and C. You will need to im-plement A as a register, which will need a reset feature. In addition, a clock or control signal is needed to tell the circuit when new values of B and C are available.

4 Most synthesis software will not generate divider circuits, unless it is a divide by a power of 2. For more general division, explicit circuitry may need to be created. The simplest division algorithm is to calculate A/B by repeatedly subtracting B from A until the result is zero or negative. The number of subtractions is then the answer. For example 7/3: 7–3 = 4 -> 4–3 = 1 -> 1–3–2. In this case the answer is 2 (the number of subtrac-tions less 1). Develop such a VHDL circuit. Try alternative methods of division, which can be found in most computer arithmetic text books.

6.6 SUMMARY

This chapter has demonstrated how VHDL can be applied in the design of datapath logic; in particular, adder and multiplier circuits. The relational and arithmetic operators that are built in to the language should always be used with care because of the amount of hardware that they can produce, especially when used inside Loop statements!

Through overloading the capability exists to modify existing subprograms and language operators. These can be replaced altogether or enhanced by adding new variants with the same name.

A number of important points about the constructs, examples and syn-thesis results in this chapter are as follows.

- Some arithmetic operators are limited to power of two operations so that hardware complexity is controlled.
- Integers cannot be accessed in a bitwise manner but the synthesizer rep-resents them internally in a binary sign-magnitude format.
- For Loops are usually supported by synthesizers, the While and Infinite Loops may not be supported. Always provide a means of exit from a loop.
- The logic described in an N-times loop will be replicated N times as the loop is unrolled.
- A function can be called as part of a sequential or concurrent assign-ment statement.
- If allowed, some operators will share hardware resources implicitly and make significant savings in area.
- The optimizer uses a number of techniques to increase the speed of a circuit, all of which can cause a significant increase in area.

7

Control logic design

7.1 BASIC PRINCIPLES

7.1.1 Which architectural style?

The simplest controller is one that begins an operation or process when it receives the on signal and stops the process when it receives the off signal. The digital equivalent is control logic that generates an output X when the control input is A and an output Y when the control input is B. VHDL can, of course, represent this operation in different ways, including using sequential statements such as **Case** or **If** and concurrent statements such as **Selected** signal assignment.

What happens, though, as the control logic becomes more complicated? For example, if the output X is dependent on inputs A, B and C, or even the inputs A, B and C and the present outputs W and Z, how is this sort of operation best represented? Storing the present state of the outputs obviously requires some form of sequential logic, and this must be implemented using a **Process** statement. It is already known that the **Case** statement generally creates a more efficient circuit than its **If-Else** equivalent. It is also a neat structure in which to construct the finite state machine that will usually be used to implement such control operations.

The issues associated with using sequential logic are as important as ever here, but constructing a finite state machines requires that wider issues are also considered. In particular the following should be taken into account.

- The initialization of the state machine. How to ensure that operation begins in a known state. This is of course related to the initialization of the storage elements.
- The stability of the state machine. Designing the circuit so that bad (unused) states are avoided and ensuring that the state of the machine can be recovered. This could mean that the circuit must be reinitialized manually or all bad states automatically recover themselves and return to a known state.

Both of these have an effect in terms of the hardware requirements of a circuit. But what happens if a design does not consider all the possible states? Although it may be sufficient to provide coverage in hardware for all input permutations of '0' and '1', an accurate simulation of the circuit must know how to deal effectively with other signal strengths such as high impedance. If the description of the circuit does not implement such measures then pre- and post-synthesis simulation differences can easily arise and the correct operation of the circuit cannot be guaranteed.

7.1.2 Multivalued logic systems

So are the two states '0' and '1' an accurate representation of the signals in a circuit? How should the designer construct a design that accurately models in simulation the occurrence of a high impedance input signal yet still synthesizes the required two-state logic operation?

There are a number of physical circuit conditions that may give rise to signals with differing strengths and values:

- three-state logic elements will produce high impedance signal strengths;
- multiple, conflicting drivers on an input may give an unknown signal value;
- the use of **resistive** pull-ups will produce weak signal strengths.

In datapath logic 'poorly defined' signals should only really occur initially and will simply flow through the circuit as it settles into a stable state. However, in control logic each storage element will prevent this settling of the circuit and there is never a guarantee that clocking the system to change its state will eventually return it to a reset state or even a stable one. Hence, the vital need for an initialization capability with all sequential logic.

The two-state logic system defines the state values of '0' and '1'. The signals are considered to be driven by a single source and are therefore of strength **forcing**. As a result, it would seem clear that this two-value logic system does not accurately represent all possible signal strengths and values that could occur in a real circuit or that must be modelled for an accurate simulation before and after synthesis.

From the designer's viewpoint this means that it is not sufficient to believe that if X does not equal '1' it automatically equals '0'. This will increase the complexity of a VHDL description, particularly affecting the structure of conditional statements such as If and Case, most commonly used in control logic design. This aspect is discussed below.

There are many multivalued logic systems available in the VHDL environment. BIT is the simplest, but the IEEE library and tool vendor libraries contain others that are in common use and so can be considered portable. The IEEE STD_LOGIC_1164 package contains the standard multivalued logic type std_logic from which many others are declared as subtypes. This is discussed below. There are a range of systems that extend the modelling capabilities of this one, containing 64 logic states or more. These are highly overspecified for synthesis and possess limited support among different vendors. They are generally most useful for performing simulation of complex bus-based or similar systems.

A number of multivalued logic systems are discussed in the rest of this section. They contain states that represent high impedance, unknown and many other signal types. Unfortunately, synthesis tool vendors have not agreed on a clear strategy for synthesizing these states. This means that synthesizers may treat them in different ways and, in some cases, may not allow a particular state to be used for some forms of expression. The guidelines that are given below for the interpretation of these states by a synthesizer are beginning to be commonly accepted as a standard.

Three-state logic system

The three-state logic system contains the values '0', '1' and 'X', which denotes a forcing but unknown logic value, not a don't care condition.

A forcing unknown state will occur in a simulation when two or more sources drive a signal simultaneously or the stable output of a flip flop has not yet been established. In hardware, the value of a signal in either of these cases will not actually be unknown but will be either '0' or '1'. However, it would not be possible to determine which it will be from any of the inputs to the circuit.

A synthesizable description cannot contain signals that are assigned an unknown value. However, tests for unknowns are allowed as part of If statements. Only the = and /= operations may be used and the synthesizer will always return the following results:

'**if** ....= 'X'...' will always return FALSE.
'**if** .../= 'X'...' will always return TRUE.

This means that the synthesizer will never produce any hardware for the case when a signal has the value 'X', but the simulator will still model the specified behaviour for such a condition.

The three-value logic system is declared as a subtype of the std_logic type and called X01. Its declaration is located in the IEEE package STD_LOGIC_1164.

Four-value logic system

This system contains the states X, 0, 1 and Z, representing the high-impedance state. It therefore typically describes signals that are buffered with high-impedance logic gates, often at an output that is controlled by some external enable signal. This multivalued logic system is also used to model a signal with multiple drives, as will be seen in the examples contained in this chapter.

Whenever a signal is assigned the value Z a three-state gate will be synthesized, as shown in the example below.

```
if SEL='I' then
    Y <= 'Z';
else
    Y <= A;
end if;
```

A signal can only be assigned a high impedance value under a controlled condition, such as inside an If or a Case statement.

A signal, such as Y, that could have been assigned a high impedance value can only be read as part of the operand in an = or /= operation. In each case the following rules apply when synthesizing:

'if ....= 'Z'...' will always return FALSE.
'if .../= 'Z'...' will always return TRUE.

So, like the unknown signal value X, a test for Z does not produce any hardware but will simulate as specified in the model.

This multivalued logic is very commonly used and is provided under various guises. The IEEE libraries contain declarations for the type Logic4 and the subtype X0IZ. Vendors usually supply arithmetic operators for the X0IZ type as well as providing vectored types containing X0IZ elements.

The standard logic system

The STD_LOGIC_1164 package contains the industry-standard set of multivalued logic systems. The standard logic system comes in two forms:

- the *unresolved* type std_ulogic;
- the *resolved* type std_logic.

Both are formed from the nine states:

1	U	Uninitialized
2	X	Forcing unknown
3	0	Forcing 0
4	1	Forcing 1
5	Z	High impedance
6	W	Weak unknown
7	L	Weak 0
8	H	Weak 1
9	–	Don't care.

These states are usually sufficient to provide an accurate model for most designs. The uninitialized state indicates that the signal has yet to be assigned a value. This is useful to know during simulation. The don't care state is particularly important for state machine design as the synthesizer can use it to reduce the logic requirements of the circuit. As before, it can be tested for in an = or /= operation and also used in any assignment statement

Note that it is illegal to mix 'U', 'Z' or '-' with any other encoding character in a string. For example:

"UUU","00X" or "---" are legal,
"UU0","XX-" or "01-" are illegal.

The difference between resolved and unresolved types is discussed in Box 7.7. Of the two types declared in the STD_LOGIC package, std_logic is the one from which a variety of subtypes are declared and sets the main standard. As already mentioned, the X01Z is a subtype of this type, but others such as UX01Z also exist. These can exploit the wide range of functions available for the type and at the same time maintain their portability.

Mixing object types in a design

The multivalued logic systems outlined above are subsets and supersets of other systems. However, as the discussion in Chapter 3 on types and subtypes explained, objects of different types cannot be universally interchanged or equated. For example, given the type declarations:

type X01Z **is** ('X','0','1','Z');
type BIT **is** ('0','1');

the conditional equation below would be illegal if A was of type BIT and B was of type X01Z.

...
if A=B **then**... -- *illegal type mismatch*

Also, it would still be illegal to reverse the equation and write 'B=A', even although A is a subset of B. Type mismatches in hardware are not a concept that is understood. However, for consistency throughout a design simulation and synthesis process, object types must be matched.

Fortunately, to enable some flexibility, a range of type conversion functions are available within standard IEEE and vendor libraries that will allow conversion between types. These can be of particular benefit when using old components or modules in a design that was created using different types. The only restriction set by many synthesizers is that type conversion cannot be done in the component instantiation statement.

If the type BIT had been declared as a subtype of X0IZ, and not a new type, the situation would have been different. The subtype declaration would have been as follows.

subtype BIT **is** X0IZ('0','I');

In this case, the test for 'A=B' would not result in a type mismatch but B can still be X or Z – states that A cannot possess. To prevent this ambiguity occurring, **strength stripper** functions can be called to convert such excess states to allowed states. See Box 7.1 for more details of these functions and where they can be used.

Synthesis issues of multivalued logic systems

It has already been stated that in a multivalued logic system it is not correct to assume that 'not 0' is 'I'. Consider the statement:

S <= A **when** C='0' **else** B;

If C is of type std_logic and its state is not '0' then B will be assigned to S. In a simulation, this action will occur not just if C='I' but also if it is 'X', 'Z', 'W', etc. However, the synthesizer will use C as the selection input for a 2-to-1 multiplexer, assuming that if C does not equal '0' then it will equal 'I', choosing the new value for S as appropriate. This is really what we wish to construct with this statement, but when a multivalued logic system is required, for implementing high impedance or don't care logic for example, simulation results before and after synthesis will differ. The designer must therefore rewrite the code to deal with the extra signal states.

If we assume that only a three-valued logic system is being used in the above example, the states '0', 'I' and 'X' could occur. The assignment statement could then be expanded to provide coverage for the extra state as follows.

S <= 'X' **when** C='X' **else**
    A **when** C='0' **else**
    B **when** C='I';

For larger value systems this construct can be expanded further. This statement will work fine for both simulation and synthesis as the synthesizer always evaluates 'C='X'' as FALSE. This will mean that the first assignment of an unknown value to S is always skipped and therefore not synthesized.

Some synthesizers may not return FALSE in the test for the unknown. To get around this the assignments could be swapped over as follows:

S <= A **when** C='0' **else**
    B **when** C='I' **else**
    'X';

Box 7.1 How to use type conversion functions and strength strippers

Type conversion functions convert a given value from its present type to a new type. These functions are necessary so that data can be passed between objects of different types while the semantics of the language are preserved. In hardware, the concept of types does not exist but the compiler will still flag type mismatch errors that occur because of their highly ambiguous nature.

As type conversion is not required at the gate level, these functions are not designed to be synthesized. If they were, a significant amount of hardware would be produced every time one was called. The vendor and standard conversion functions are not automatically translated into hardware. This is achieved through the use of the Built_in attribute discussed in Chapter 6.

The IEEE library supplies a wide range of type conversion functions. For example:

function TO_BIT (s :std_ulogic) **return** BIT;

converts a std_ulogic value to one of type BIT. The name of all conversion functions take this 'To_xxxx' form.

Also provided are functions called strength strippers. These convert a value of a certain type to a value of a particular subtype of that type. Hence, they strip out a number of possible states. For example,

function TO_X0IZ (s : std_ulogic_vector) **return** std_ulogic_vector;

would convert any bits of an arbitrary length vector with std_ulogic elements to a same-length vector containing only 'X', '0', 'I' or 'Z'. Any of the other four states that may occur with this data type ('U', 'W', 'L', 'H' or '-') would be converted to 'X'. Strength strippers are useful for cutting down on the number of possible combinations that must be covered by an If statement, say when only a certain input combination will affect the result. Note that the returned type is identical to that of the parameter – the function converts the *value stored in the object* to one of the possible subtype values but does not convert the object itself.

Both conversion functions and strength strippers can be used like normal functions. The only exception to this is that some synthesizers may not allow them to be used in parameter lists when calling other functions or procedures. Additionally, neither function can be called in the port map of a component instantiation statement.

Here no comparison is made with an unknown and the synthesizable conditions are covered before the default assignment is reached. This is therefore ignored by the synthesizer. The same principles can be applied with a Case or If statement inside a process as in the following example.

```
case C is
    when '0' => S <= A;
    when '1' => S <= B;
    when others=> S <='X';
end case;
```

However, some synthesizers may also throw out the assignment of the unknown to S even though it will be ignored. In this case it may be necessary to resort to conditional compilation of the design.

Conditional compilation allows the designer to force the synthesizer to skip any non-synthesizable parts of a design, effectively switching off the compiler. Non-synthesizable constructs have not been discussed in any detail in this text. Specifically, they are the language elements that are not supported by the synthesizer as opposed to those that are just ignored, such as initial values on signals. Non-supported constructs cause a synthesis error; ignored ones will only generate a warning.

The example below demonstrates how the conditional compilation switch can be used to make the synthesizer ignore the assignment of an unknown to A.

```
signal C : std_logic;
...
process (..., C, Y, ...)
begin
    ...
    if C='0' then
        A <=Y;        -- known signal value
    elsif C='1' then
        A <= not Y;   -- known signal value
    -- compile off
    else
        A <= 'X';     -- unknown forcing value
    -- compile on
    end if;
    ...
end process;
```

The compile off/on commands are synthesizer specific and so the reader should refer to his or her synthesis manual.

Using the switch extensively throughout a description can make it very difficult to read and it is not specifically provided to make a complex simulation also synthesizable. It is better to create two separate files in this case.

The compile off/on command also allows other non-synthesizable constructs to be implemented.

- TheTextio package can be used so that important simulation messages and warnings can be relayed to the designer.
- The file type can be used so that simulation data can be stored to disk in a user-defined format.

7.2 SEQUENCE FOLLOWER

This section presents an example that demonstrates some of the more advanced topics in VHDL. These not only include language constructs but also techniques that are useful for design state machines and control logic in general. The aspects of sequential logic design with VHDL that were introduced in Chapter 5 are assumed to be understood at this stage. Topics covered in this section include:

- concurrent statements
      Block
- attributes
- multivalued logic
- the resolution of signals with multiple drivers.

  Instead of presenting a wide variety of different architectures, as in the previous three chapters, a more detailed example is given, which presents the new topics using only two designs.

7.2.1 Functional overview

A sequence follower circuit watches the sequence of numbers it receives and indicates, through an output signal, when a particular sequence has been received. This example applies this principle to a particular application – a keypad-operated door lock.

The circuit to be designed must watch for a particular sequence of four digits occurring on the keypad. This keypad contains the numbers 0 to 6 along with a CLEAR button, which acts as a system reset. When a digit is entered it is converted into a binary format and passed to the sequence follower circuit. If the digit sequence 3625 is entered then a signal is output to light an LED and another is output to the unlock mechanism. Whenever the door is physically open a signal is sent back to the sequence follower. All external signals are active low.

The design has been decomposed into two parts, as shown in Figure 7.1. Both contain independent state machines. The decoder receives as input the binary-encoded keypad digit on A, B and C, along with the signal OPENED, which indicates that the door is physically open. For each correct digit received in the sequence, a signal S is sent to the enabler. If the sequence is broken or the clear button is pressed, a reset signal R is sent to the enabler. If the door is open, the decoder is disabled and a reset pulse is sent.

After the enabler has received a reset pulse it will wait to receive four sequence pulses on S from the decoder. If it receives these, it unlocks the door and lights the LED. If a reset pulse is received at any time the counter is reset.

The state diagram of the decoder circuit is shown in Figure 7.2. There are five states. The coding scheme associated with each transition shows the inputs required to make the transition and the outputs that transition produces. The inputs to the decoder are latched and so it stays in each state until the input changes but will only issue a pulse to the encoder when it first enters the state. The next correct number makes the transition to the next number state whereas an incorrect number will reset the system. The CLEAR button on the keypad is encoded with the binary value I I I and in this case can simply be treated as an incorrect input.

Once the final correct number has been entered the system will wait in that state until the door has been opened. This is the reset trigger.

The state diagram of the enabler is shown in Figure 7.3. This has the same number of states as the decoder. From the reset state called LOCK, each sequence pulse that is received causes a state transition. If a reset pulse is received at any time the system is reset. The inputs S and R can never go low together.

The outputs from the enabler do not change until the transition is made into the Unlock state. In this state the LED remains lit and the door unlocked until the enabler receives a reset signal from the decoder. This will only be generated by the decoder when the door is opened.

The state machines in each block are fully synchronous, rising edge-triggered circuits. Each is driven by one phase of a two-phase non-overlap-

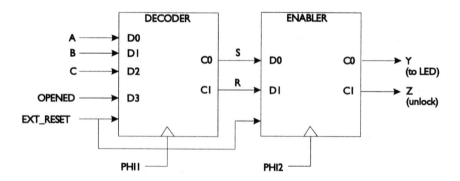

Figure 7.1 The two elements of the sequence follower circuit.

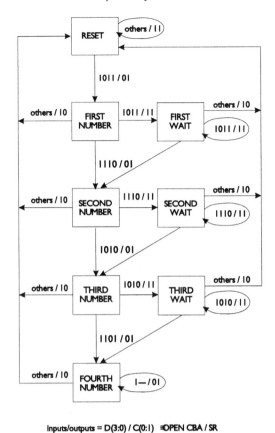

Inputs/outputs = D(3:0) / C(0:1) ≡ OPEN CBA / SR

Figure 7.2 State diagram for the Decoder block when EXT_RESET='1'.

ping clock. Although it would be simpler to construct a single-state machine to implement this circuit, two have been used to demonstrate different VHDL principles.

A second version of the system will also be constructed. This contains a manual override input that unlocks the door and disables input from the keypad.

7.2.2 VHDL code design

Library selection

To demonstrate the use of multivalued logic systems the industry-standard std_logic type is used throughout this example. This will have conse-

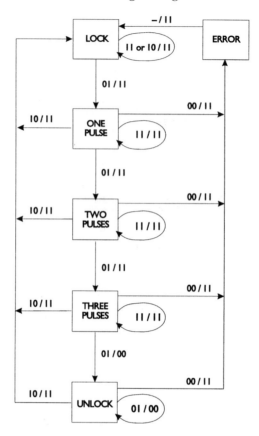

inputs/outputs = D0 DI / CO CI= SR / YZ

Figure 7.3 The state diagram for the Enabler block when EXT_RESET='I'.

quences for the way in which the code is designed, particularly in terms of the coverage of states in the finite state machine.

The std_logic type is the resolved form of the std_ulogic type and is well supported by the functions and operators declared in the IEEE library package STD_LOGIC_1164. The std_logic_vector is the array form of the type. This package must therefore be visible to each design file in the sequence follower.

The entity declaration

The top level of the proposed circuit was shown in Figure 7.1. The two blocks, Decoder and Enabler, have clear functions and well-defined I/O. The design process may therefore begin at this level. As a result, the top-level entity of the sequence follower contains a Port statement with data andclock inputs **A, B, C, OPENED, PHI1** and **PHI2** and the control outputs **Y** and **Z**.

The architecture body

One way to implement this circuit would be to use a structural style and instantiate components that implement the behaviour of the **Decoder** and the **Enabler**. It is not possible to use a function in this case as:

- present state values must be stored between calls but a function cannot retain information;
- sequential logic cannot be generated in a function as it cannot contain **Wait** statements or signal assignments.

Thus, the **Decoder** has been implemented as a component. Figure 7.4 shows the VHDL description of this circuit, which demonstrates how to construct a state machine. It is always advisable to use enumerate type encoding for states as shown in the architecture's declarative part. This example has eight states, inferring three storage elements and using all possible permutations of '0' and '1'. The present state is stored on a signal **PRESENTSTATE** that is visible only to this architecture.

The architecture contains two processes and implements a Moore state machine, dependent only on the present state for its outputs. The combinational logic that determines the outputs is in the process **OUTPUT_SELECT**. Of course, it would also have been possible to implement this logic concurrently in a **Conditional** or **Selected** signal assignment statement. By using a process, the sequential logic of state selection and the combinational logic of output selection are completely separated and can be independently, and therefore more easily, optimized.

The process **STATE_SELECT** has the sole task of selecting the new state on each clock cycle. The external reset signal is first tested before the nested **Case** statements evaluate the new state from the present state and the inputs. As with examples in previous chapters, the external signal is not directly given a new value, but is assigned to a variable called **NEXTSTATE**. After the new state has been determined the final statement in the process then reassigns the variable to the signal.

So let us look at the **Case** constructs. The three bits used to encode **PRESENTSTATE** provide eight logic permutations for the synthesizer to consider. There are eight states examined by the first **Case** statement and so, as far as the synthesizer is concerned, a final **Others** clause is not required. On the other hand, the simulator has not been given an initial value for **PRESENTSTATE**, which would be ignored by the synthesizer anyway. If a signal or variable is not given an initial value for simulation before it is read, it automatically assumes one. See Box 7.2 for details of how it selects this value. A signal of type std_logic will be given an initial value of 'U', so **PRESENTSTATE** will have the value 'UUU'. To cover this and any other val-

```
library IEEE;
use IEEE.std_logic_1164.all;
entity DECODER is
    port (   signal DIN : in std_logic_vector(3 downto 0);
             signal CK, CLR : in std_logic;
             signal COUT : out std_logic_vector(0 to 1) );
end DECODER;

architecture BEHAVIOUR of DECODER is
    type DECODERSTATES is    (RESET, FIRST_NUM, FIRST_WAIT,
                             SECOND_NUM, SECOND_WAIT,
                             THIRD_NUM, THIRD_WAIT,
FOURTH_NUM);
    signal PRESENTSTATE : DECODERSTATES;
begin
    OUTPUT_SELECT : process(PRESENTSTATE)
    begin
      case PRESENTSTATE is
        when RESET => COUT <= "10";
        when FIRST_NUM | SECOND_NUM |
            THIRD_NUM | FOURTH_NUM => COUT <= "01";
        when others => COUT <= "11";
      end case;
    end process;

    STATE_SELECT : process
      variable NEXTSTATE : DECODERSTATES;
    begin
      wait until (CK'event and CK='1');
      NEXTSTATE := PRESENTSTATE;
      if CLR='0' then
        NEXTSTATE := RESET;
      else
        case PRESENTSTATE is
          when RESET =>
            case DIN is
              when "1011" => NEXTSTATE := FIRST_NUM;
              when others => null;
            end case;
          when FIRST_NUM =>
            case DIN is
              when "1110" => NEXTSTATE := SECOND_NUM;
              when "1011" => NEXTSTATE := FIRST_WAIT;
              when others => NEXTSTATE := RESET;
            end case;                          -- continued...
```

Figure 7.4 Description of the Decoder component in the sequence follower.

```
-- Decoder architecture continued...
            when FIRST_WAIT =>
                case DIN is
                    when "1110" => NEXTSTATE := SECOND_NUM;
                    when "1011" => null;
                    when others => NEXTSTATE := RESET;
                end case;
            when SECOND_NUM =>
                case DIN is
                    when "1010" => NEXTSTATE := THIRD_NUM;
                    when "1110" => NEXTSTATE := SECOND_WAIT;
                    when others => NEXTSTATE := RESET;
                end case;
            when SECOND_WAIT =>
                case DIN is
                    when "1010" => NEXTSTATE := THIRD_NUM;
                    when "1110" => null;
                    when others => NEXTSTATE := RESET;
                end case;
            when THIRD_NUM =>
                case DIN is
                    when "1101" => NEXTSTATE := FOURTH_NUM;
                    when "1010" => NEXTSTATE := THIRD_WAIT;
                    when others => NEXTSTATE := RESET;
                end case;
            when THIRD_WAIT =>
                case DIN is
                    when "1101" => NEXTSTATE := FOURTH_NUM;
                    when "1010" => null;
                    when others => NEXTSTATE := RESET;
                end case;
            when FOURTH_NUM =>
                case DIN(0) is
                    when '0' =>      NEXTSTATE := RESET;
                    when others => null;
                end case;
            when others => null;
        end case;
    end if;
    PRESENTSTATE <= NEXTSTATE;
end process;

end BEHAVIOUR;
```

Figure 7.4 Description of the Decoder block (continued).

ues of std_logic, such as 'X', the Others clause is added. It has no function but is simply included for completeness.

The important feature of this code is that there is an external reset for the state machine. Thus, it is not really important what happens to PRESENTSTATE or any other signals before the reset is applied. It is assumed that, once in a stable state, the system will take care of itself. When the logic tests the value of DIN in a certain state, the hardware will default to the state indicated in the Others clause if none of the specified bit patterns occur. Remember from Chapter 4 that the hardware produced will not explicitly test for all bit permutations covered by this clause. Instead, it relies on the fact that all other explicit tests have failed. Any signals that are of high impedance, for example, will therefore be caught by this implicit test. In a simulation, the Others clause will also cover the occurrence of other values on DIN, such as 'U' or 'X'.

The Decoder state machine has been a relatively straightforward design. The Enabler's has fewer states but requires more thought. First, rather than using a component instantiation, the Enabler has been constructed inside a Block statement. This is a concurrent statement that allows a design to be partitioned hierarchically by creating an internal architecture block. For synthesis this statement is useful for organizing and grouping the operations in the architecture. Effectively it can be thought of as an architecture within an architecture, and in a simulation may even contain a Port statement that defines the I/O of the block. Only the external signals that this contains are then visible inside the block. Unlike a process, a block can contain both variable and signal declarations in its declarative part. These are local to the block. Also, its statement part contains concurrent not sequential statements. In this respect it is similar to a component but it cannot be duplicated as easily. More specific details of the construct are given in Box 7.3 and the description of the sequence follower circuit that contains the Enabler block and the Decoder component is given in Figure 7.5.

Box 7.2 How to simulate without using initial values

Initial values are always ignored by the synthesizer but not by the simulator. If a signal or a variable is not given an initial value when it is declared then when the description is compiled for simulation, it automatically assumes one. For an enumeration type, this value will be the left-most emuneration literal in the list given in the type's declaration. For example, a signal of type std_logic will be given a value 'U' and one of type BIT will assume a value of '0'. The same applies for numeric types such as integer. An integer object with a range −10 to 10 will assume a value of −10. If it had a range of 3 downto −3, the initial value would be 3.

The Enabler block groups together its incoming data and outgoing control signals with a record. The signal I/O is declared of this record type in the block's declarative part and therefore is only visible inside the block. The mapping of the block's I/O signals occurs in the concurrent signal assignment statements. These illustrate how Selected names can be used to access individual elements of array types contained in the record. The benefit of using a record with a Block statement in this manner is simply that in the absence of a Port statement the record allows the designer to clarify the

Box 7.3 How to use the Block statement

1 The Block statement is a concurrent statement that defines an internal block in a design. Its use is allowed in synthesis without full support for all its capabilities. Apply the syntax:

```
block_statement ::=
  label: block [ is ]
         block_declarative_part
  begin
         concurrent_statements
  end block [ label ];
```

2 The label is compulsory.
3 For synthesis, a block simply allows a group of concurrent statements to be grouped together in a particular architecture. When designing for simulation it is possible to create guarded blocks. Any **guarded signal assignments** within a **guarded block** will only drive the signal when the **guarded expression** for the block is TRUE.
4 Like an entity, a block may declare a header part that contains Port and Generic statements. The port map controls the visibility of signals in the block from the higher level architecture. Neither the Port nor Generic statements are generally supported by synthesizers within a Block statement.
5 The block declarative part may contain declarations for:

> subprograms, types, subtypes, constants, signals,
> shared variables, components, attributes.

It may also contain subprogram bodies, attribute specifications and Use clauses. Some other declarations are not supported for synthesis.
6 Any concurrent statement may be used inside a block. This allows a design to be decomposed by the hierarchical nesting of blocks.
7 The hierarchy in a description that is created by a Block statement will usually be flattened by the synthesizer. An attribute such as Merge can often be used to preserve this hierarchy (Box 7.6).

```
library IEEE;
use IEEE.std_logic_1164.all;
use WORK.STATE_DEFN.all;
entity SEQUENCE is
    port (   A, B, C, OPENED, EXT_RESET : in std_logic;
             PHI1, PHI2 : in X01Z;
             Y, Z : out std_logic );
end SEQUENCE;
architecture BEHAVIOUR of SEQUENCE is
    component DECODER
        port (   signal DIN : in std_logic_vector(3 downto 0);
                 signal CK, CLR : in std_logic;
                 signal COUT : out std_logic_vector(0 to 1) );
    end component;
    signal S,R : std_logic;
    signal DEC_DATA : std_logic_vector(3 downto 0);
    signal DEC_CTRL : std_logic_vector(0 to 1);
begin
    DEC_DATA <= A & B & C & OPENED;
    S <= DEC_CTRL(0);
    R <= DEC_CTRL(1);
    DECODE : DECODER port map ( DIN=> DEC_DATA, CLR=>
                    EXT_RESET, CK=> PHI1, COUT=> DEC_CTRL );
    ENABLER : block
        type ENABLERIO is record
            D : std_logic_vector(0 to 1);
            C : std_logic_vector(0 to 1);
        end record;
        signal IO : ENABLERIO;
        signal PRESENTSTATE : ENABLERSTATES;
    begin
        IO.D(0)<=S;
        IO.D(1)<=R;
        Y<=IO.C(0);
        Z<=IO.C(1);
        OUTPUT_SELECT : process(PRESENTSTATE)
        begin
            case PRESENTSTATE is
                when UNLOCK => IO.C <= "10";
                when others =>    IO.C <= "01";
            end case;
        end process;                          -- continued...
```

Figure 7.5 Description of the sequence follower circuit containing the **Decoder** component and the **Enabler** block.

```
-- Enabler block continued...
    STATE_SELECT : process
      variable NEXTSTATE : ENABLERSTATES;
    begin
      wait until (PHI2='1');
      NEXTSTATE := PRESENTSTATE;
      if EXT_RESET='0' then
        NEXTSTATE := LOCK;
      else
        case PRESENTSTATE is
          when LOCK =>  case IO.D is
                  when "01" =>       NEXTSTATE := ONE;
                  when "11"|"10" =>  null;
                  when others =>     NEXTSTATE := ERROR;
                    end case;
          when ONE =>    case IO.D is
                  when "01" =>    NEXTSTATE := TWO;
                  when "11" =>    null;
                  when "10" =>    NEXTSTATE := LOCK;
                  when others =>  NEXTSTATE := ERROR;
                  end case;
          when TWO =>   case IO.D is
                  when "01" =>    NEXTSTATE := THREE;
                  when "11" =>    null;
                  when "10" =>    NEXTSTATE := LOCK;
                  when others =>  NEXTSTATE := ERROR;
                  end case;
          when THREE => case IO.D is
                  when "01" =>    NEXTSTATE := UNLOCK;
                  when "11" =>    null;
                  when "10" =>    NEXTSTATE := LOCK;
                  when others =>  NEXTSTATE := ERROR;
                  end case;
          when UNLOCK =>case IO.D is
                  when "01"|"11" =>  null;
                  when "10" =>       NEXTSTATE := LOCK;
                  when others =>     NEXTSTATE := ERROR;
                  end case;
          when ERROR =>    NEXTSTATE := LOCK;
        end case;
      end if;
      PRESENTSTATE <= NEXTSTATE;
    end process;
  end block ENABLER;
end BEHAVIOUR;
```

Figure 7.5 Description of the sequence follower (continued).

structure of the block's I/O. Manipulation of records and record elements is discussed in Box 7.4.

The declarative part of the block contains a declaration for the signal PRESENTSTATE. This is local to the block, and the same signal may exist in another block, component or subprogram in the same architecture without problems. If it also existed in the architecture or a higher level block that contains this block, then the declaration in the block would be ignored.

The signal PRESENTSTATE is of type ENABLERSTATES. This type is not declared in the architecture but is located in the package STATE_DEFN in the working library. Figure 7.6 shows the package and its contents. ENABLERSTATES is an enumeration type containing six values. Five of these represent the states of the Enabler, as shown in Figure 7.3, and the sixth, ERROR, is used to handle illegal input conditions. In the real hardware the only illegal input condition will be '00'. If this occurs, all we want the hardware to do is to return to a reset (locked) state rather than entering an error state first. When simulating, the illegal inputs may include the occurrence of unknowns or uninitialized signals. To allow debugging, the designer may want to indicate that an error has occurred by actually entering the error state.

This can be achieved through the use of user-defined enumerate type encoding as the package illustrates. The attribute ENUM_TYPE_ENCODING allows the binary encoding of an enumerate type to be specified by the user. In this case, the binary encoding of the five states in the state machine is the same as would be generated automatically. The final state, however, is represented by the binary pattern '000'. This is the same as the LOCK state, which indicates that whenever the ERROR state is entered the hardware should return directly to the LOCK state. As far as the hardware is concerned therefore, the LOCK and ERROR states are the same state, so why does the Case statement inside the STATE_SELECT process look for the condition of "...PRESENTSTATE=ERROR..."? This happens because the ENUM_TYPE_ENCODING attribute is synthesis specific and therefore is completely ignored by the simulator. During simulation therefore, ERROR is considered another state and will cause the pre- and post-synthesis simulations to differ. The synthesizer will ignore this additional state.

Despite the simulation differences caused by user-defined encoding, it is a useful technique that allows the designer to match the enumeration literals to the most meaningful binary representations. It is only one of the synthesis-specific attributes that tool vendors provide. Most of these relate to timing constraints on signals or area constraints on sections of a design. These attributes are passed on directly to the logic optimizer.

Usually, supported attributes are provided in a package. This will contain the Attribute Declaration statement for each attribute. Having made this package visible through a Use clause, Attribute Specification statements associate or **decorate** an item with the attribute and value. For example, if the

Box 7.4 How to access and assign record values

The elements of a record must be accessed using a selected name. A **selected name** references an element of a record by expanding the name. Consider the type declaration below.

```
type BUS_STRUCTURE is record
    address : BIT_VECTOR(15 downto 0);
    data : BIT_VECTOR(7 downto 0);
    read, dtack : BIT;
end record BUS_STRUCTURE;
```

A signal *a_bus* declared as this type could have values assigned to different elements as follows:

```
a_bus.address <= X"FE80";   -- all elements of a composite record element
a_bus.address(15) <= '0';   -- an indexed element of a composite record
element
a_bus.data(7 downto 4) <= "1100" -- a slice of elements of a composite
record element
a_bus.read <= '1';          -- a scalar record element
```

When using elements of records in expressions, selected names are also used. Note that aggregate assignment must be used to assign all elements of a record at once. Named associations can be used in such an assignment as normal. See the examples below.

```
a_bus.address <= X"00" & a_bus.data; -- record element in an expression
a_bus <= ( X"34F4", "0000_1010", '1', '0'); -- aggregate positional association
a_bus <= ( data => '0', address => X"1FFF", others => '0' );
                        -- aggregate named association
```

```
package STATE_DEFN is
    attribute ENUM_TYPE_ENCODING: STRING;
    type ENABLERSTATES is (LOCK, ONE, TWO, THREE, UNLOCK,
                            ERROR);
    attribute ENUM_TYPE_ENCODING of ENABLERSTATES: type is
                            "000 001 010 011 100 000";
end STATE_DEFN;
```

Figure 7.6 Package for the user-defined enumerate type and encoding scheme.

maximum area that the sequence follower must occupy is 60 000 square microns then the statement

attribute MAX_AREA **of** SEQUENCE : **entity is** 60,000;

could be added to the entity's declarative part. The use of other user-defined attributes for synthesis is often limited, but declaring and specifying attributes are discussed further in Box 7.5.

Both the attribute declaration and specification for ENUM_TYPE_ENCODING are used in the package STATE_DEFN. They are placed in a package in the first place because some synthesizers will not allow attribute declarations anywhere except in a package or an entity. More details of the synthesis attributes discussed above and others are contained in Box 7.6.

The operations of the block are very similar to the component. The process OUTPUT_SELECT determines the output required for the present state, whereas STATE_SELECT performs the appropriate state transition on the second clock, PHI2. For synthesis purposes, the Others clause in this Case statement covers the occurrence of the input condition '1 1'. When simulating, it will also cover all other values as already mentioned. The ERROR state is ignored by the synthesizer as it is encoded with the same binary pattern as the LOCK state.

Let us create another version of the sequence follower circuit. Often, such a keypad door entry system will contain a manual override operated by a key. When the key is turned to the on position the door is permanently unlocked and the keypad will no longer function. There are a number of ways to implement such an addition to the system.

- Add the override signal to the logic of the state machines in a similar way to that used to implement the external reset.
- Drive the outputs directly, independent of the state of the rest of the system.

This second option requires the same output signals to be driven by multiple sources, as Figure 7.7 shows. Multiple drivers for a signal can be generated by two concurrent statements, such as two processes or just two concurrent signal assignment statements. In the circuit shown in Figure 7.7, the two signals are connected through high-impedance buffers. This is only one method: Wired And or Wired Or logic can also be used. This design infers high-impedance buffers, but the designer can also specify the method through compile time options. The changes and additions to the sequence follower architecture SEQUENCE that are required for the manual override operation are shown in Figure 7.8. A new block called MANUAL_OVERRIDE has been added to implement the required logic. The OVERRIDE signal (or its inverse) will be used to drive the enable inputs on the buffers as appropriate.

Attribute declaration

An **attribute declaration** is used to create a user-defined attribute. Synthesis tool vendors use attribute declarations to create attributes that pass information from the design units to the synthesizer, optimizer, etc. (see Box 7.6). As user-defined attributes will not be understood by these tools, their use is limited and often they are not supported for synthesis. An attribute declaration has the syntax:

```
attribute_declaration ::=
    attribute attribute_identifier : type_name|subtype_name;
```

1 The attribute_identifier is the name or **designator** of the attribute.
2 The type or subtype of the attribute could be anything from a record to a constrained INTEGER.

Attribute specification

An **attribute specification** associates a user- or vendor-defined attribute (not the predefined attributes discussed in Chapter 6) with one or more items from a particular class and gives the attribute an appropriate value. It has the following syntax:

```
attribute_specification ::=
    attribute attribute_identifier of item_name {, item_name} : item_class is
                                                            expression;

item_class ::=
    entity    | architecture   | procedure   | function   | package
    | type    | subtype        | constant    | signal     | variable
    | component | label         | literal     | units
```

1 When an item is given an attribute value in a specification it is said to be **decorated**.
2 If an attribute is being attached to a package, the item name would be the name of the package. Alternatively, if it is being attached to a component, it is the instance name of the component. Other examples are shown in the accompanying text.
3 The item name can be replaced by the keywords All or Others. The former will decorate all items in the specified class with the expression. The latter will assign the expression to all items in the class that have not yet been decorated with this attribute.
4 An attribute specification *must* be located in the same declarative part as the declaration of the item to which it refers. The attribute declaration itself must only be visible to the specification and may therefore be local or placed in a package
5 The expression must be of a type or subtype appropriate to the attribute declaration.

Box 7.6 How to use synthesis specific attributes

Tool vendors usually provide a number of additional attributes that enable the designer to interface more easily with the synthesis and optimization environment. These attributes allow information to be placed inside the description that relates to how a design or any of its labelled items, such as a process, should be compiled or optimized. Assigning an attribute with a **specification** is discussed in Box 7.5. The following are typical examples of the attribute declarations provided by synthesis tool vendors. Most of these are tool vendor-specific so the reader should check his or her own documentation.

Timing/area constraining attributes

    attribute MAX_AREA : REAL;   --max. desired area, in unit area.
    attribute MAX_TRANSISTION : REAL; --max. signal transistion time
    attribute MAX_DELAY : REAL;   --max. delay from an input signal
                                        change to a change in an output.
    attribute LOAD : REAL;   --the load on an output port, in unit load

Structural constraining attribute

    attribute MERGE : BOOLEAN;

This attribute determines whether the hierarchy of a design unit or of a concurrent statement (such as a process) within an architecture will be preserved during compilation. By default, the hierarchy of a **Block** statement is collapsed by the synthesizer but that of a process is preserved. To preserve the block an attribute specification could be used as follows:

    attribute MERGE **of** block_name : **label is** FALSE;

Enumeration encoding attribute

Tool vendors usually allow the design to specify a user-defined encoding scheme for enumeration types. The default encoding scheme assigns the value 0 to the first enumeration literal, 1 to the second, and so on. A user-defined scheme allows the designer to specify the number of bits needed to represent each literal and the value each literal is represented by. This can simplify the hardware requirements and produce a more meaningful binary representation of a machine's states. The ENUM_TYPE_ENCODING attribute (or similar) is provided for this purpose. For example, the following specification assigns a new encoding scheme to a three-element enumeration type called PRIMARY.

    attribute ENUM_TYPE_ENCODING : STRING;
    type primary **is** (RED, BLUE, GREEN);
    attribute ENUM_TYPE_ENCODING **of** PRIMARY : **type is** "001 010 100";

As far as the synthesized circuit is concerned, wiring together the outputs from the high-impedance buffers is a perfectly correct implementation. Assuming that the logic for the buffer enable inputs is valid, the output will be driven high or low by only one source. However, how does a simulation know the required strength and value of such a signal? One statement says that Y should be 'I', for example, while the other says it should be 'Z'. The signal must therefore be resolved using a **resolution function**. A resolution function is automatically called every time an assignment is made to a signal with multiple drivers so that its new value can be determined. Only signals declared to be of resolved types such as std_logic or its subtype X0IZ should be used for signals with multiple drivers. See Box 7.7 for more details on resolution functions.

7.2.3 Synthesis and optimization

Preliminaries

Through a behavioural-style architectural description, it is relatively easy to implement a complex control task with VHDL. However, it is good practice for the designer to break down the complete function into smaller routines for reasons of code complexity and readability. The synthesizer will generate a similar circuit whether it is written as a single routine or many smaller routines. The physical difference will be the hierarchy created in the latter case. However, using smaller sections of code also has many advantages.

- The designer can see more clearly the effect of particular language constructs on the circuit produced. The design process is therefore an important learning process.

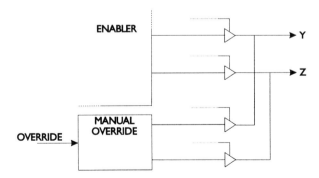

Figure 7.7 Sequence follower circuit with multiple drivers.

Box 7.7 How to handle signals with multiple drivers

A signal has multiple drivers when:

- it is assigned values in more than one process within the same architecture; or
- it is assigned values in more than one concurrent signal assignment statement.

A signal with multiple drivers must be of a type that can be resolved.

Synthesis of signals with multiple drivers

When synthesized, multiple drivers for one signal are connected together in one of three ways:

- through wired Or gate(s).
- through wired And gate(s).
- through high-impedance buffers. The enable input for each buffer is derived by the conditional expression that selects that particular output. If a design specifies the use of high-impedance buffers by actually assigning 'Z' to a signal then this option *must* be selected. Otherwise the outputs of the buffers will be wired together through And or Or gates and not simply connected.

Simulation of signals with multiple drivers

The simulator must be able to determine an appropriate value for a signal W when being driven by the two concurrent statements:

```
W <= S;
W <= T;
```

First, W must be a resolved type. That is, its type declaration is linked to a resolution function that is called every time an assignment is made to the signal. This function examines the strength and value of each driver and determines an appropriate new signal strength and value. For example, if S='X' and T='0', then W would be assigned the value 'X', as the unknown value in S makes it impossible to resolve the two signals. However, if S='Z' and T='0' then W will be '0'. The type used throughout this chapter, std_logic, is the resolved version of the std_ulogic type. A subtype of a resolved type, such as X01Z, is also resolved.

The synthesizer will ignore any resolution function associated with a type and connect the sources together in the manner discussed above.

-- No changes to the Library or Use clauses

```
entity SEQUENCE is
    port (    A, B, C, OPENED, EXT_RESET : in std_logic;
              PHI1, PHI2, OVERRIDE : in X01Z;
              Y, Z : out std_logic );
end SEQUENCE;
architecture BEHAVIOUR of SEQUENCE is

    ...
    -- No change to architecture's declarative part

    ...
begin

    ...
    -- No change to concurrent signal assignments or Decoder instantiaition

    ...
    ENABLER : block

        ...
        -- No change to the Enabler block declarations

        ...
    begin
        IO.D(0)<=S;
        IO.D(1)<=R;
        Y<=IO.C(0) when OVERRIDE='0' else 'Z';
        Z<=IO.C(1) when OVERRIDE='0' else 'Z';

        ...
        -- No change to the OUTPUT_SELECT process
        -- No change to the STATE_SELECT process

        ...
    end block ENABLER;
    MANUAL_OVERRIDE : block
    begin
        Y <= '1' when (OVERRIDE='1') else 'Z';
        Z <= '0' when (OVERRIDE='1') else 'Z';
    end block MANUAL_OVERRIDE;
end BEHAVIOUR;
```

Figure 7.8 Modifications (shown in bold) required for the sequence follower entity and architecture to implement the manual override option.

- Sections of a design (such as components) are easier to test as they can be simulated in isolation.
- Optimization can take place at several levels. Optimizing each small section of a design independently is faster than trying to perform a glo-bal optimization on the whole design. Furthermore, if some parts need to be redesigned, synthesis and optimization can target just those ele-

Control logic design

ments. Random control logic is the type of logic that most often benefits from area optimization. However, the size of the block of logic being optimized is influential in the effectiveness of the process. As a result, there needs to be a trade-off between optimization time (related to the size of the circuit) and the quality of the optimization.

The effect on circuit area of flattening a design during optimization as opposed to preserving its hierarchy has been demonstrated in earlier examples. Here, the structure of the control logic is considered the important feature of the design and therefore will be retained. Unlike the datapath logic in Chapter 6, timing constraints on control logic are often less critical. Therefore, we will concentrate on minimizing the area in the optimization process.

In the multiple driver example, we are interested in how the synthesizer will deal with the two drivers of the output signals. We have inferred high-impedance drivers for the two output signals and must therefore ensure that the compile time option for selecting the connection scheme – wired logic type – is set to tristate (TRI). See the Dos and Don'ts section for more information on selecting the correct wired logic type.

Remember that the package STATE_DEFN must be compiled and stored in the working library before SEQUENCE-BEHAVIOUR is synthesized.

Logic synthesis and optimization results

This circuit consists of one component that describes the function of the Decoder and a block that describes the Enabler. Both these contain two processes that select the next state and the new output. The top level of the circuit that results from synthesis is shown in Figure 7.9. As the Decoder is a component, it has appeared as a hierarchical block (the decoder entity and architecture are synthesized before the host entity and architecture, SEQUENCE-BEHAVIOUR, can be synthesized). The Block statement that contained the description of the enabler's function has not appear as a hierarchical block however. The block has been removed and the two process blocks are placed on the top level. Effectively, the block was never there. This has occurred because by default the synthesizer collapses the hierarchy of a Block statement but preserves that of a Process statement. The merge attribute can be used with particular instances of blocks or processes to change the default, discussed earlier in Box 7.6.

Figure 7.10 shows the next level of the hierarchy below the decoder block. This contains the two blocks that represent this block's Process statements. The statistics for the synthesized circuit contained in all four of these processes are shown in Table 7.1. As the Block statement has been flattened by the synthesizer, the two processes in the SEQUENCE-BEHAVIOUR entity-architecture were originally contained in the Enabler block. The circuits associated with these statistics are not shown. As can be observed, the proc-

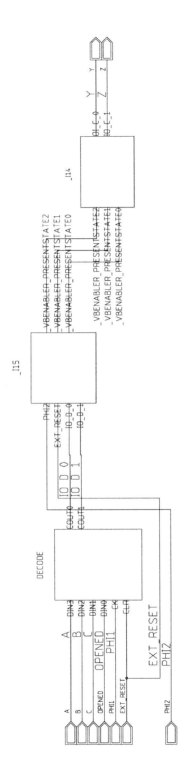

Figure 7.9 Top-level synthesized block for the sequence follower.

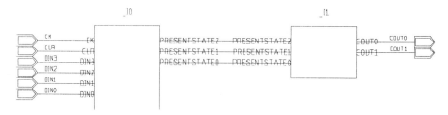

Figure 7.10 Top level of the synthesized Decoder component.

esses containing the state selection routines – the actual state machines – are much larger than the output selection processes. Of the two, the Decoder's state machine is the larger.

So let us see how these four processes optimize. In each case, low, medium and high levels of synthesis effort were applied and the results at each stage are shown in Table 7.2.

The OUTPUT_SELECT process in the Enabler block must produce a default output value for all states except one. Hence, the simple logic has benefited only slightly from optimization. The circuit changes have involved the removal of redundant and duplicated logic, which has naturally resulted in a reduction in the circuit's delay. The initial size of the circuit has

Table 7.1 Logic synthesis statistics for the processes in the Decoder component (DECODER-BEHAVIOUR) and Enabler block (SEQUENCE-BEHAVIOUR)

Parameters *Temperature: 25.0 °C* *Voltage: 5.0 V*	*Synthesis* *Library: generic*	
Circuit: SEQUENCE-BEHAVIOUR (Enabler block)	OUTPUT_SELECT process	STATE_SELECT process
Cells	4	89
Transistors	12	462
Area	1760	57 917
Longest latch to latch delay/ max. frequency	n/a	10.020 ns/99 MHz
Longest input to output delay	2.869 ns	n/a
Circuit: DECODER-BEHAVIOUR	OUTPUT_SELECT process	STATE_SELECT process
Cells	23	116
Transistors	102	588
Area	13 026	73 760
Longest latch to latch delay/ max. frequency	n/a	10.588 ns/94 MHz
Longest input to output delay	7.470 ns	n/a

resulted in a minimized circuit with only a medium level of effort. The optimized circuit is shown in Figure 7.11.

The equivalent process in the **Decoder** component implements a more complex logic function, but the encoding scheme that has been used in this case has enabled a significant reduction in logic requirements when optimized. The states are encoded as illustrated in the truth table shown in Table 7.3. In this case, the default encoding scheme has enabled the optimizer to minimize the circuit to just use an **Or** gate and **Inverter**, as shown in Figure 7.12. Here, it is unlikely that a user-defined encoding scheme could have improved on this.

The two state selection processes have optimized with different levels of success. The **STATE_SELECT** process in the **Enabler** block has reduced its transistor count almost fourfold, whereas the similar process in the **Decoder** component has reduced the number of transistors by less than three-fold. Furthermore, the latter process required all three levels of effort compared with only one for the former – no change occurs to the circuit for medium and high levels of effort. There are two main differences between these processes.

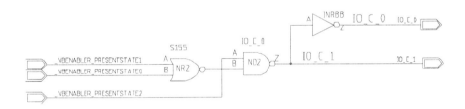

Figure 7.11 Optimized OUTPUT_SELECT process in the **Enabler** block.

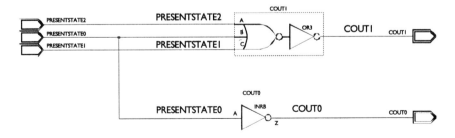

Figure 7.12 The optimized OUTPUT_SELECT process in the **Decoder** component.

Table 7.2 Logic optimization statistics for the processes in the Decoder component (DECODER-BEHAVIOUR) and Enabler block (SEQUENCE-BEHAVIOUR)

Parameters *Temperature: 25.0 °C* *Voltage: 5.0 V*	*Optimization* *Type: area* *Level: Low*	*Medium*	*High*
Circuit: SEQUENCE-BEHAVIOUR OUTPUT_SELECT process			
Cells	2	2	No change
Transistors	10	8	No change
Area	1232	1056	No change
Longest input to output delay	1.868 ns	1.466 ns	No change
STATE_SELECT process			
Cells	15	No change	No change
Transistors	132	No change	No change
Area	15843	No change	No change
Longest latch to latch delay/ max. frequency	9.470 ns /105 MHz	No change	No change
Circuit: DECODER-BEHAVIOUR OUTPUT_SELECT process			
Cells	5	2	No change
Transistors	22	10	No change
Area	2816	1232	No change
Longest input to output delay	5.731 ns	3.123	No change
STATE_SELECT process			
Cells	36	35	26
Transistors	268	260	204
Area	31687	30807	24293
Longest latch to latch delay/ max. frequency	11.615 ns /86 MHz	11.596 ns /86 MHz	11.268 ns /88 MHz

Table 7.3 Truth table for the output assignments in the OUTPUT_SELECT process of the Decoder component.

State, present state	*State encoding*	*Output, COUT(0 to 1)*
RESET	000	10
FIRST_NUM	001	01
FIRST_WAIT	010	11
SECOND_NUM	011	01
SECOND_WAIT	100	11
THIRD_NUM	101	01
THIRD_WAIT	110	11
FOURTH_NUM	111	01

- The Decoder's state selection process must deal with all eight possible states, whereas the Enabler's has only to deal with five. The unused states in the Enabler can be employed in logic minimization.
- The Enabler must decode only two inputs compared with the Decoder's four. Although the proportion of possible states that are explicitly decoded by the Enabler is greater (3 out of 4 compared with 2 out of 16), the logic is still simpler.

It is therefore clear that the numbers of states and inputs have influenced not only the size of the circuit, but also the complexity of the optimization process. This has occurred despite the similarity of the processes' structure. The circuit for the Enabler block's STATE_SELECT process is shown in Figure 7.13 and for the STATE_SELECT process in the Decoder in Figure 7.14. Each of these circuits consists of a block of logic driving three flip flops that store the current state. The outputs of the flip flops drive the outputs of the block as well as being fed back into the circuit, where they are used to determine the next required state.

The second form of this circuit contains a manual override, and the synthesis result for this version is shown in 7.15. As this figure shows, three-state buffers have been used and the two drivers for each signal Y and Z have been connected together. The MANUAL_OVERRIDE block has been flattened by the synthesizer but consists of an inverter for the OVERRIDE signal and another that drives low the outputs when OVERRIDE is high. When OVERRIDE is low the main circuit buffers are enabled and the override buffers are disabled. The contents of each block in this circuit are unchanged from the previous version of the circuit and so no optimization has been demonstrated.

7.2.4 Sequence follower summary

The sequence follower circuit demonstrated in this section illustrated the use of the Block statement in synthesis. Its usefulness in this situation is limited to the simple segregation of code sections rather than allowing an 'architecture within architecture' hierarchical style of design.

The example has also highlighted other 'designing for synthesis' rather than 'designing for simulation' differences. The designer can create user-defined attributes, but these are generally not supported for synthesis. However, a number of synthesis-specific attributes, such as ENUM_TYPE_ENCODING, are available. This one is particularly valuable for state machine design as it can enable the encoding of states to be simplified.

The second part of this example also demonstrated how signals can be driven by multiple sources. Again, the approach required to achieve this when designing for synthesis is different to that required for a simulation-

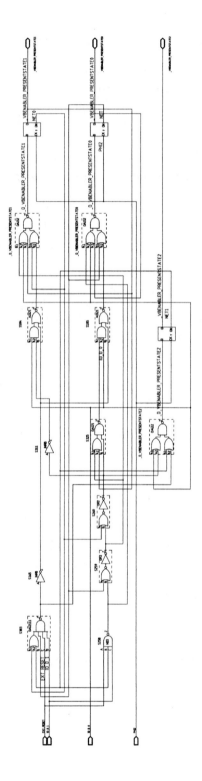

Figure 7.13 Optimized STATE_SELECT process for the Decoder.

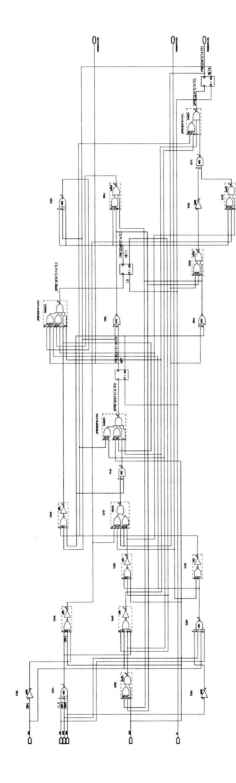

Figure 7.14 STATE_SELECT process in the Decoder.

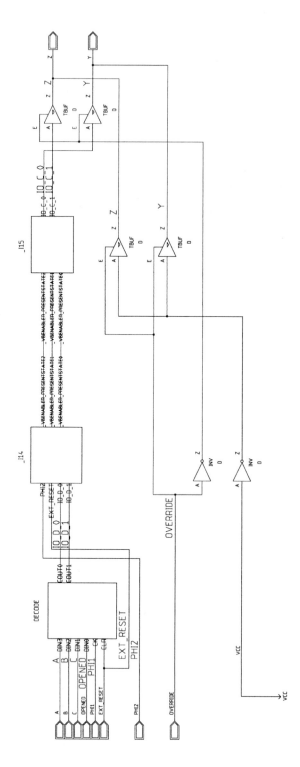

Figure 7.15 Sequence follower with a manual override using three state gates.

only design. The resolution functions used in the latter to determine the new value and strength of a signal with multiple drives are ignored when synthesizing. Instead, the usual approach is to connect the signal sources together through three-state buffers.

7.3 RANDOM SEQUENCE GENERATOR

A random sequence generator starts from a 'seed' and produces a new sequence of output bits whenever it receives a clock pulse. It is only a pseudorandom sequence however, as the same seed will always produce the same following sequence.

This section demonstrates the use of only one new statement – the procedure – in the design of two forms of the generator. It also shows new important examples of statements and important approaches demonstrated or discussed in previous sections. These include:

- the use of Inout ports;
- how unconstrained objects can be used to create flexible circuit modules;
- three-state buffers;
- the placing of declarations in the entity declarative part;
- more multivalued logic.

The Procedure statement is also special in that it may be used in both concurrent and sequential sections of code. The two examples below will demonstrate this.

7.3.1 Functional overview

Two high-level circuit diagrams are shown in Figure 7.16. Figure 7.16a illustrates the simplest circuit in which the inputs are a clock signal and a reset. The output is shown in this case to be 4 bits. Whenever the reset is low the sequence is reset to 0001. After reset is released, each clock pulse well generate a new output pattern of bits.

Internally, the random sequence generator operates by shifting all the bits one position to the right and creating the new left bit through an Xor function on two of the other bits. This will be demonstrated clearly in the VHDL.

To combat the problem of the same seed always generating the same pseudorandom sequence, the design in Figure 7.16b can be loaded with a new seed. Externally, the structure looks identical to the circuit in Figure 7.16a The key difference in this case is that the output ports for the bit pattern are now bidirectional as they are used to load the new seed. This has significant consequences for the way in which these ports are designed as the circuit cannot drive the external lines when a new seed is being loaded.

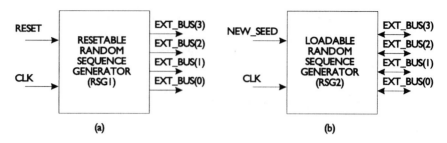

Figure 7.16 The inputs and outputs from the top-level blocks in the two variants of the random sequence generator

This calls for the use of three-state buffers, encountered in the last section. The following functionality is required.

- When the new seed signal is high, it is assumed that there is no external seed signal driving the input of the circuit. (This cannot be controlled from within this circuit as it is a function of the external circuit providing the new seed patterns.) A new output pattern will be generated for each clock pulse.
- When the new seed signal is low, the circuit must not drive the external ports as a new seed pattern is being loaded. When the new seed returns to the high state, operation continues as before.

7.3.2 VHDL code design

Selecting libraries and packages

Four-state X0IZ logic is demonstrated in this example. As a resolved subtype of the std_ulogic type it is defined in the STD_LOGIC_1164 package. A vector of this type is not defined in the package and so a new package – LOGIC_UTILS – has been included, which defines the X0IZ_VECTOR and the associated logical operator functions. This package is actually supplied by the CAD tool vendor and is therefore specific to the toolset used in this book. However, X0IZ is a common object type and all vendors will provide libraries to support it. See Appendix B for the description of the package.

The procedure that describes the operation of the random sequence generator is stored in a package called CONTROL_LOGIC in the working library. The details of the procedure itself are discussed below.

The entity

The entities for the two designs are named RSG1 and RSG2. As the top-level circuit designs showed in Figure 7.16, both have a similar port structure. RSG1 has RESET and CLK inputs and the random sequence generated on the bus EXT_BUS. RSG2 differs in that it has a NEW_SEED input rather than a reset, which is used for loading the sequence's new start pattern. To load the pattern, EXT_BUS is bidirectional and defined as type Inout. This allows the port to be assigned or read within the architecture body. In previous examples it has always been necessary to assign a signal of type Out to a variable before it could be read within the circuit.

Architecture design

RSG1-BEHAVIOUR is the entity–architecture pair for the simple random sequence generator circuit. It is shown in Figure 7.17.

One of the key features of this code is its modularity. The bit width of the sequence generator is only specified in the port declaration statement of the entity. In this case, it is 4 bits wide, but it could have been any value from 3 upwards. This lower limit is set by the procedure RND_GEN and will be explained shortly.

Building components that are modular in this way are obviously of great benefit to design time. Using procedures that are modularized can cut down on the number of discrete elements in a design and so assist in the management of large designs. The modular approach has been possible through the use of attributes such as **Range** rather than explicit parameters to specify the length of a vector.

The design shows a behavioural architecture contained within a process. The random pattern is stored in the register RANDOM_REG, which is assigned a value at the end of the process. VRANDOM_REG is used throughout the sequential section to select whether the new random pattern will be generated from the existing pattern or the reset pattern '1000'''. To generate this reset pattern, a For Loop is used with the iterations determined by the Range attribute. This allows a pattern of any length to be generated. The If statement selects the left bit and sets it to '1'. Remember that the For Loop is a sequential statement, unlike the For-Generate, which can be used in the concurrent statement part.

The procedure that generates the new random pattern is RND_GEN, which is called from within the process. The actual procedure is located in the package CONTROL_LOGIC shown in Figure 7.18. Like function declarations, the procedure is split into two parts:

- the declaration contained in the package declarative part;
- the body located in the package body.

In fact, many of the general rules for functions can also be applied to procedures. The key difference is that any of the objects passed in the pa-

```
library IEEE,DZX;
use IEEE.STD_LOGIC_1164.all;      -- Declaration for X01Z
use DZX.LOGIC_UTILS.all;          -- Declarations for X01Z_VECTOR
use work.CONTROL_LOGIC.RND_GEN;  -- Random sequence generator
entity RSG1 is
   port (   CLK, RESET : in X01Z;
            EXT_BUS : out X01Z_VECTOR(3 downto 0));
end RSG1;
architecture BEHAVIOUR of RSG1 is
   signal RANDOM_REG : X01Z_VECTOR(EXT_BUS'range);
begin

   process
      variable VRANDOM_REG : X01Z_VECTOR(RANDOM_REG'range);
   begin
      wait until (CLK='0' and CLK'event);
      if RESET='1' then
         VRANDOM_REG := RANDOM_REG;    -- use old pattern
      else
         for K in RANDOM_REG'range loop      -- use "1000.."
            if K=RANDOM_REG'left then
               VRANDOM_REG(K) := '1';
            else
               VRANDOM_REG(K) := '0';
            end if;
         end loop;
      end if;
      RND_GEN(VRANDOM_REG);               -- generate new pattern
      RANDOM_REG <= VRANDOM_REG;
   end process;

   EXT_BUS <= RANDOM_REG;

end BEHAVIOUR;
```

Figure 7.17 Simple random sequence generator with reset.

rameter list can be modified and returned to the calling architecture. Each parameter can be passed as an explicit type such as signal or variable. Alternatively, as this example shows, the object will select a default type depending upon the mode of the port. In this procedure, RANDOM defaults to a variable because it is a bidirectional port. Within the procedure this variable can be read or modified. The rules for passing and returning sig-

```
library IEEE,DZX;
use IEEE.STD_LOGIC_1164.all;
use DZX.LOGIC_UTILS.all;
package CONTROL_LOGIC is
   procedure RND_GEN (RANDOM : inout X01Z_VECTOR);
end CONTROL_LOGIC;
package body CONTROL_LOGIC is
   procedure RND_GEN (RANDOM : inout X01Z_VECTOR) is
   variable NEW_BIT : X01Z;
   begin
      NEW_BIT :=  RANDOM(0) xor RANDOM(2);
      RANDOM := NEW_BIT & RANDOM(RANDOM'left downto 1);
   end RND_GEN;
end CONTROL_LOGIC;
```

Figure 7.18 CONTROL_LOGIC package containing random sequence generator.

nals and variables and default object types are explained in more detail in Box 7.8.

The procedure itself is defined within the package body. The statements it contains are executed sequentially, like a process. RND_GEN is a modular procedure that can generate a new random pattern of any length greater than 3 bits. This minimum is determined by the Xor logic that defines NEW_BIT using the current state of the RANDOM variable. This assignment takes immediate effect because NEW_BIT is a variable. The new value of RANDOM is also assigned immediately in the next statement. If RANDOM had been passed to this procedure as a signal then the assignment would not take place until the end of the calling process.

In this context, a procedure can therefore be thought of as a subprocess or module that can be called many times by the same or different processes. It is called sequentially. See Box 7.9 for more details.

Figure 7.19 shows the entity–architecture pair RSG2-BEHAVIOUR, which demonstrates the other way in which a procedure can be used – as a concurrent statement. In this example, the procedure IOBUFFER is used to create three-state input and output buffers. Both calls to this procedure are made concurrently. The sequential call to the RND_GEN procedure still exists within the process GENERATOR.

This behavioural description requires the buffers contained within the BUFFERS block to enable and disable the input and output ports connecting the random sequence generator to the external bus. The code demonstrates another way of declaring and defining a procedure – both the procedure declaration and body are contained within the entity declarative part.

Box 7.8 How to declare a Procedure

A procedure is a subprogram that can receive and return any number of values. Like a function, it can consist of both a separate declaration and a body. When storing the procedure in a package, the procedure declaration is placed in the package declaration and the procedure body is stored in the package body. The syntax for both statements is:

procedure_declaration ::= **procedure** simple_name [(parameter_list)];
procedure_body ::=
    procedure simple_name [(parameter_list)] **is**
      procedure_declarative_part
    begin
      sequential_statements
    end [**procedure**] [simple_name];

1 In the parameter list:

- values can be passed to a procedure using objects of mode In;
- values can be passed to a procedure and the updated values returned to the calling design unit using objects of mode Inout;
- values can be returned from a procedure using objects of mode Out.

See Box 7.9 for details of how a procedure will modify an object.
2 The parameter list may contain signals, variables and constants. See Box 7.9 for details of the class of objects that can be passed to each class of objects in the parameter list.
3 The following assumptions are made for the class of an object in the parameter list when it is unspecified:

- any object of mode In is treated as a constant;
- any object of mode Inout or Out is treated as a variable.

4 The declarations allowed in the declarative part are the same as those in a function declarative part. They are all reset between calls to the procedure and any initial values will be ignored by the synthesizer.
5 A Wait statement cannot be used inside a synthesizable procedure.

The IOBUFFER procedure itself is relatively simple. The parameter list uses default object types such that EN and A are treated as constants and Y is treated as a variable. The use of the range attribute to determine the width of the buffer makes this procedure modular and portable. Note that EXT_BUS appears in both instances of the procedure call – once as an input and once as an output. This is a valid use of EXT_BUS, which is defined in the entity as mode Inout.

Figure 7.20 illustrates more clearly the function of this circuit for the nth bit of an M-bit pattern (where n<M). The state of the input and output buffers is determined by the value of NEW_SEED. This signal is also used to determine the state of the RANDOM(n) – the bit that will be fed into the REG_GEN procedure. This selection operation is defined in the logic of the synchronous process contained in this architecture. The second block of logic in the diagram defines the shift operation for this bit of the random pattern. The result, the bit in the new random sequence is latched by the flip flop.

This example may seem a little complicated, but it clearly demonstrates how a procedure can be used in both a sequential and concurrent manner. The concurrent procedure is effectively a process. This process may contain a Clock statement (but not a Wait statement; Box 7.8) or may simply be a block of combinational logic, as in this case. As with any concurrent statement, variables cannot be passed to the procedure but can be used inside it, as seen here. Refer again to Box 7.9 for more details.

7.3.3 Synthesis and optimization

The synthesis and optimization results for these circuits are mainly of interest for two reasons:

- the fact that preset/clear inputs are not used on the flip flops in RSG1;
- the demonstration of three-state logic implementation in RSG2.

The statistics for the two circuits are presented, but only two circuit diagrams are illustrated.

Synthesis and optimization results

The synthesized circuit of **RSG1-BEHAVIOUR** is not shown here, but it is easy to visualize its structure – a top-level block representing the process and a four-stage sequential logic circuit. The procedure is not created as a separate block but is incorporated into the logic of the process. Using a Block statement around the procedure and specifying that it should not be flattened during synthesis would be one way of preserving its hierarchy.

Again, the synthesized circuit for **RGS2-BEHAVIOUR** is not illustrated. The **IOBUFFER** block implements the three-state buffers and the **GENERA-TOR** process defines a similar function to its equivalent in **RSG1-BEHAV-IOUR**. The key difference is that this circuit can be reloaded and the other can only be reset.

Table 7.4 shows the statistics for both the synthesized and optimized versions of **RSG1-BEHAVIOUR** and **RSG2-BEHAVIOUR**. The statistics for the process and the block in the latter circuit are given separately as both were

Box 7.9 How to call a Procedure

A procedure is a statement and therefore, unlike a function, it is not called as part of an expression. Furthermore, it can be treated as both a sequential statement and a concurrent statement. The same procedure can be used in both ways, assuming a concurrent procedure returns and receives signals and not variables. The behaviour of the procedure is always executed in a sequential manner. The simple syntax of a procedure call is:

procedure_call ::= simple_name [(parameter_list)]

The following rules must be applied to the passing of parameters to different classes of object in a procedure:

- a signal can only be passed a signal;
- a variable can only be passed a variable;
- a constant can be passed an expression containing signals, variables, constants and / or numeric literals.

Positional and named association can both be used in a procedure call.

Sequential procedure call

A sequential procedure call is one made within any sequential section of a design unit, such as process. All classes of object can be passed to the procedure. Assignments that are made in the procedure to variables passed with modes Inout or Out will take immediate effect, but a value assigned to a signal will not be established until the end of the sequential statement section. Hence, assignments within the procedure follow the same rules as assignments made in the calling sequential section itself. A sequentially called procedure cannot be used to infer sequential logic.

Concurrent procedure call

A concurrent procedure call is one made within any concurrent section of a design unit. Variables cannot be passed to a concurrent procedure unless they are shared and declared in the declarative part of the calling design unit. Assignments that are made in the procedure to objects passed with modes Inout or Out take immediate effect on the associated objects in the calling design unit. Hence, assignments within the procedure follow the same rules as assignments made in the calling concurrent section itself.

A concurrent procedure is called whenever its Inout or Out parameters change. It is therefore like a commonly used process that can be called from a package whenever it is required.

```
library IEEE,DZX;
use IEEE.STD_LOGIC_1164.all;
use DZX.LOGIC_UTILS.all;          -- Declarations for X01Z_VECTOR
use work.CONTROL_LOGIC.RND_GEN;  -- Random sequence generator
entity RSG2 is
   port (   CLK, NEW_SEED : in X01Z;
            EXT_BUS : inout X01Z_VECTOR(3 downto 0));
procedure IOBUFFER ( signal EN : in X01Z;
                     signal A : in X01Z_VECTOR;
                     signal Y : out X01Z_VECTOR);
procedure IOBUFFER ( signal EN : in X01Z;
                     signal A : in X01Z_VECTOR;
                     signal Y : out X01Z_VECTOR) is
begin
   for K in X01Z_VECTOR'range loop    -- sequential loop statement
      if EN='1' then Y(K) := A(K);
      else        Y(K) := 'Z';
      end if;
   end loop;
end;

end RSG2;

architecture BEHAVIOUR of RSG2 is
   signal RANDOM_REG : X01Z_VECTOR(3 downto 0);
   signal OLD_SEED : X01Z;
begin
   BUFFERS : block
   begin
      OLD_SEED <= not NEW_SEED;
      IOBUFFER (NEW_SEED, EXT_BUS, NEW_REG_VALUE); -- input buffer
      IOBUFFER (OLD_SEED, RANDOM_REG, EXT_BUS); -- output buffer
   end block BUFFERS;
   GENERATOR : process
      variable VRANDOM_REG : X01Z_VECTOR(RANDOM_REG'range);
   begin
      wait until (CLK='0' and CLK'event);
      if NEW_SEED='1' then    VRANDOM_REG := RANDOM_REG;
      else                    VRANDOM_REG := NEW_REG_VALUE;
      end if;
      procedure RND_GEN(VRANDOM_REG);
      RANDOM_REG <= VRANDOM_REG;
   end process;
end BEHAVIOUR;
```

Figure 7.19 Random sequence generator with external seed.

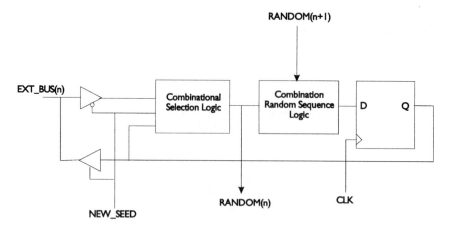

Figure 7.20 The functional blocks in the architecture RSG2-BEHAVIOUR.

optimized independently. Synthesis flattens the block structure as it was not tagged to be preserved.

The optimized circuit for RSG1-BEHAVIOUR shows very little change from the original synthesized version. Only during the high level of effort did the area reduce. The optimized circuit is shown in Figure 7.21. The two main characteristics of this circuit are as follows.

- The combinational logic between each flip flop is limited to one or two gates. This has restricted the effectiveness of the optimization process. A more complex logic structure between each pair of storage elements would have enabled more logic permutations to be investigated by the optimizer.
- The RESET signal is synchronous and is part of the combinational logic input to each flip flop. The circuit behaviour could have been specified as asynchronous by performing the reset sequence outside the clocked part of the process. This would have inferred flip flops with Preset and Clear inputs. The actual construction of a behavioural description to produce this circuit is left for the reader to attempt. Beware, other changes will also be required to the process!

The optimized top-level circuit for RSG2-BEHAVIOUR is shown in Figure 7.22. The block represents the optimized process, which is not shown in detail. As the statistics for the process indicate, it has a similar area to RSG1-BEHAVIOUR and, in this case, area optimization has had no effect.

The circuit shows the four input and four output three-state buffers attached to EXT_BUS. The enable inputs for four of these are fed from NEW_SEED. The other four are fed from the same but inverted input signal. The state of NEW_SEED therefore determines the function of the bus.

Table 7.4 Logic synthesis and area optimization statistics for the RSG1-BEHAVIOUR and RSG2-BEHAVIOUR circuits

Parameters Temperature: 25.0 °C Voltage: 5.0 V	Synthesis Library: generic	Optimization Type: area Level: Low	Medium	High
Circuit: RSG1-BEHAVIOUR				
Cells	12	No change	No change	10
Transistors	106	No change	No change	102
Area	13555	No change	No change	12850
Longest input to latch delay	2.634 ns	No change	No change	2.216 ns
Circuit: RSG2-BEHAVIOUR GENERATOR process				
Cells	14	No change	No change	No change
Transistors	122	No change	No change	No change
Area	15 315	No change	No change	No change
Longest input to latch delay	3.913 ns	No change	No change	No change
BUFFER block				
Cells	37	9	No change	No change
Transistors	138	82	No change	No change
Area	20 068	10 210	No change	No change
Longest input to latch delay	8.043 ns	6.072	No change	No change

- When NEW_SEED is low EXT_BUS is an input bus. The left bank of buffers is disabled to disconnect the output drive from EXT_BUS. The right bank is enabled and the incoming value on EXT_BUS is passed through to NEW_REG_VALUE.
- When NEW_SEED is high EXT_BUS drives the output. The right bank of buffers is disabled and the value on RANDOM_REG is passed through to EXT_BUS.

Optimization has achieved a 50% reduction in area for the **BUFFER** block. This has occurred through the removal of three inverters driving three of the output buffers. The remaining inverter now has a fan-out of four. This is a parameter that could have been specified in the synthesizer to limit the number of inputs that one output can drive and vice versa. Each component in a vendor's library will also have fan-out and fan-in parameters associated with it. The synthesizer should be able to check if a design violates either of these. If the fan-out restriction on the inverter was two in this circuit, then two inverters would be required to drive the buffer enable signals.

Finally, note the cost in delay of using the three-state buffers. A large proportion of the 8-ns input to latch delay for the synthesized **BUFFER** block is due to a single three-state buffer.

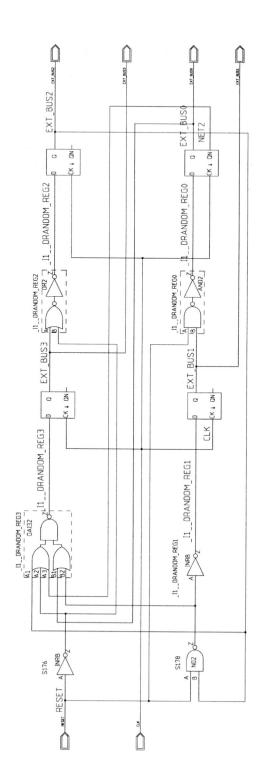

Figure 7.21 Optimized circuit for RSG1-BEHAVIOUR.

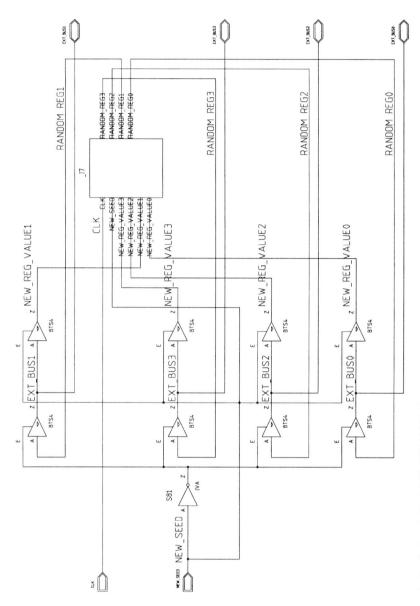

Figure 7.22 The optimized circuit for RSG2-BEHAVIOUR.

7.3.4 Random sequence generator summary

The random sequence generator has particularly demonstrated the use of the procedure as both a concurrent and sequential statement. The function itself is quite simple. A loadable seed value has enabled the use of three-state buffers plus the effect of a bidirectional signal to be demonstrated. The optimization statistics clearly highlighted the effect that these buffers have on the performance of the circuit.

Designing procedures that are modular is the key to a good design. The RND_GEN procedure used here can be used again in another design without recompiling its package. The use of attributes and non-defined vector ranges facilitates this modular approach.

7.4 DOS AND DON'TS

7.4.1 Enumerate type encoding

Synthesis allows the use of enumerate type encoding to enable logic of different states to be specified in the design. For example, states **THREE** and **FOUR** in a state machine may be encoded as '00110' and '110'. Encoding of this type is fine for synthesis but often is not understood by the simulator. This can lead to differences in simulation results before and after synthesis as, before synthesis, the simulation will ignore the new encoding scheme. When the circuit has been synthesized, the encoding scheme is contained within the logic of the circuit.

The designer must be aware of and account for these differences. Encoding schemes that are cosmetic and do not have any benefits in terms of logic reduction, for example, should be avoided.

7.4.2 State coverage for multivalued logic using If and Case statements

The topic of state coverage was discussed in detail in section 7.1. It is repeated in brief here in order to clarify the key issues.

Simulation models the function of a circuit through a description of its internal states, inputs and outputs. To provide a complete analysis, all states declared as part of the type must be included. Consider an outline architecture description containing a **Case** statement.

```
signal THIS_STATE : bit
process
begin
    case THIS_STATE is
        when '0'=> THIS_STATE <=...
        when '1'=> THIS_STATE <=...
    end case;
```

end process;

Although the logic of this statement is correct, it by no means covers all possible states that can occur during simulation. These include uninitialized, unknown or weak low to name but a few.

Synthesis converts this design into interconnected circuit elements that form the logic of the function. That means that, as the states '0' and 'I' are perfectly synthesizeable, simulation results before and after synthesis will be the same.

If a more comprehensive simulation is to be performed then the type bit is inadequate and X0IZ or std_logic types must be used. The consequences of this are twofold.

- The VHDL description must account for non-synthesizable states. This relates particularly to the use of Case and If statements, and methods of dealing with these states were discussed in section 7.1. The most extreme scenario is to switch the synthesizer off and on in different sections of the description.
- The synthesizer will ignore the actions associated with certain states and so pre- and post-synthesis differences will result in the simulation.

Do remember the key points about multivalued logic design:

- a Case statement must always provide full state coverage, even if this means gathering all remaining states under the Others statement;
- the following rules are applied to the use of enumeration literals of multivalued logic
  - An unknown, don't care or high impedance value used as an operand in an '=' ('/=') operation will always return FALSE (TRUE).
  - An unknown can only be used as an operand of an = or /= operation in an If statement. Any assignment of an unknown will generate an error.
  - A don't care can be used in an assignment. The synthesizer will automatically select the most suitable value.
  - A high impedance value can be assigned within or used in the conditional logic of a Case or If statement.

7.4.3 Restrictions on the elements of interface lists

The parameters contained in a port map, procedure, component instantiation, etc. must follow certain rules. The key restrictions are as follows.
- A clock signal must be passed to a procedure as a signal. This must be specified in the interface list as otherwise it will default to a constant.
- A type conversion or any other function is not supported in a component instantiation statement. Function calls are supported when nested in other functions or procedures.

- When mixing named and positional association in an interface list the positional association must come first. The function calls below illustrates this.

xyz (clk, a_in, d=>outbus, c=>b_in); -- legal
xyz (a => clk, a_in, b_in, outbus); -- illegal

Some of these rules are imposed because of synthesis.

7.4.4 Restrictions on procedures for synthesis

Procedures can be called as concurrent or sequential statements. Only sequential statements can be used within the procedure: it cannot contain Wait statements. To infer a clocked procedure an If statement can be used, but this must be done with caution.

- A concurrent procedure call is equivalent to a process and may contain a Clock statement.
- A sequential procedure is contained within a process. It may only contain a Clock statement if the process itself does not contain one.

Remember that multiple calls to the same procedure will result in logic being generated each time. For a concurrent procedure, multiple calls can be made through a For-Generate statement. For a sequential procedure, multiple calls can be made with a For-Loop statement.

7.5 EXERCISES

1 Return to Zero coding is a widely used telecommunications signalling protocol. It uses a spacer or empty 'E' token, which is transmitted between every data value. For example the data sequence 0,1,1,0,1 would be transmitted as 0.E.1.E.1.E.0.E.1.E. Two bits per value are needed. Data '0' is coded as 01, Data '1' as 10 and E as 00. In this way every signal involves a 1-bit change. Design a circuit that takes as input a series of binary numbers on the 1-bit input line DATA and outputs the 2-bit sequences necessary on the 1-bit signal lines X and Y.

2 Many security systems have programmable door locks. A typical 4-bit code is input one digit at a time, and if correct the door is opened. After three failed attempts, a light 'ALARM' is lit and the door is permanently locked tight. Design such a system with the following specification.

CODE: a 4-bit input signal with values 0–9 and Clear; 0–9 are numeric values. Clear is used to indicate a mistake and resets the input. A sequence of four numbers is needed

SUPERVISOR: a 1-bit input signal. If it is 1 then the code entered is the new access code. If the ALARM signal is on, then it too is reset to zero.

OPEN: a 1-bit output signal which opens the door.

ALARM: a 1-bit output signal that is set high if three sets of invalid codes are input.
This is a reasonably complex design and is best approached by breaking the problem into pieces and tackling each piece separately.

7.6 SUMMARY

This chapter has demonstrated how VHDL can be applied to the design of control logic to construct state machines with complex state maps. The main issue that has been highlighted is the use of multivalued logic in a design and how this affects both the simulation and synthesis processes.

The sequence follower example architectures demonstrated a good approach to the design of state logic using a nested case structure. The random sequence generator used a module-based design to illustrate the benefit of procedures (or even functions) that are designed to be portable.

Some of the key points about VHDL design that have emerged from the chapter include the following.

- Multivalued logic allows the function of the design to be verified for a range of possible signal strengths and values. The synthesizer does not understand 'uninitialized' or other similar signals. This can have two consequences as follows:
  - the designer must structure code such that the logic is correct and can be interpreted by both the simulator and the synthesizer;
  - simulations carried out before synthesis may generate different results to simulations carried out afterwards.
- The Block statement allows a design to be sectioned into self-contained elements. When simulating a design, these blocks can contain inputs and outputs similar to the entity. For synthesis, a block only helps to segregate the design description into readable sections. If a merge attribute is attached to the block its hierarchy will be maintained during synthesis.
- Signals can have multiple drivers within a circuit. If the design infers that three-state buffers are to be used, then the outputs can be wired together. If these are not specified then the synthesizer will use either wired AND or OR logic, depending on the parameter settings at the time of synthesis.

- Signals declared as bidirectional by using the mode Inout can be read or updated within an architecture without the need for temporary variables. The circuits produced by both approaches will be identical.
- Procedures can be passed signals, variables and constants, although strict rules apply for mixing types. Unless an object is specifically declared as a signal, it will be treated as a constant (if it has mode In) or as a variable (if it has mode Inout or Out).
- Synthesizeable procedures cannot contain Wait statements.

8
Case study

8.1 BACKGROUND

The previous chapters in this book have concentrated on particular aspects of digital electronic systems design. Of course, in any practical system a whole range of design styles and constructs will be applied. This chapter shows how the techniques introduced can be used together to develop a reasonably complex digital electronic circuit.

This chapter is of the 'putting it all together' variety. No new techniques are introduced; rather we assume that the reader has some simple experience with the design of circuits using VHDL and hence we focus on the appropriate use of these techniques.

The first part of the chapter introduces the case study. We focus on the design of a parallel processing computer targeted for the rapid execution of neural networks. We have chosen this design for a number of reasons.

- This is a real design, which we have undertaken ourselves as part of a UK research project aimed at understanding how best to construct electronics for real-time neural networks. As such, we have access to detailed design descriptions and performance data.
- Being developed at a university, no proprietary information is involved, and hence a detailed disclosure of results is possible.
- The design fulfils our criteria of being complex enough to show all the principles involved, yet is simple enough to understand and also to ensure short VHDL code fragments.

8.2 NEURAL NETWORKS

Neural networks are a way of processing information based on or inspired by the principles of computation within the human brain. Although the first pioneering work on neural networks occurred in the 1950s, from the early 1960s until the early 1980s relatively little work was performed in this area. The improvements in electronics technology enabling faster computation, allied to a strong demand for smart systems, resulted in a renaissance in research in this area. In the last 10 years, neural networks have become an enormously popular topic for research. In the last few years, the progress made has moved out of research laboratories and into the commercial market. Today, a surprisingly large number of software packages and consumer products, including washing machines and microwave ovens, use neural networks.

Conventional computer systems based on the Von Neuman concept are extremely good at executing algorithms that have been precisely formulated for them. However, there are tasks, such as image and speech recognition, that cannot be suitably or easily captured using an algorithm.

Biological neural networks can, on the other hand, perform these tasks apparently effortlessly using myriads of nerve cells (neurons). Current estimates place the number of neurons in the human brain at 10^{11}. They are organized in a complex unmapped interconnection structure in which each neuron may be connected via variably weighted links (synapses) to several thousand other neurons. The information 'learnt' by the network is stored in these synapses and processed collectively by the whole net. This densely interconnected structure allows the brain to explore many competing hypotheses simultaneously.

These biological neural models have inspired researchers to develop artificial counterparts, albeit on a smaller scale, but using the same massively parallel highly interconnected structure. Like their biological versions, artificial neural nets are capable of learning from a given set of examples, generalizing from inputs with some degree of insensitivity to minor variations from the exemplar set and abstracting essential information from data sources containing noise or redundancy. The neural net requires no programming; using only a learning law it is able to change its synaptic weights, thereby allowing the network to adapt itself to the system inputs. Artificial neural nets are also very robust; their densely interconnected and highly parallel topology makes them immune to small computation errors or the loss of small numbers of connections.

Some of the first and possibly best-known contributions were those of McCulloch and Pitts in 1943, who showed how neural-like networks could compute, and of D.O. Hebb, who in 1949 proposed a learning law (Hebb's law). These works became the basis for artificial neural network training algorithms. In the 1950s and 1960s there was a flurry of activity by researchers using single-layer networks called perceptrons. Disappointment followed, however, when the perceptrons failed to solve many simple problems. In 1960, Minsky and Papert showed by mathematical induction that the networks then in use were theoretically incapable of solving many simple problems, including the function of the simple exclusive-or gate. Since then, however, new, more powerful, learning algorithms have been produced that do not suffer the same criticisms applied to earlier works.

Now, after a period of nearly two decades, research into artificial neural networks has once more captured the interest of researchers intrigued by the potential offered by this concept. There have been many impressive demonstrations of neural networks. Networks have been trained to convert between written text and speech, recognize handwritten text and identify images, with superior performance to that shown by more traditional techniques.

The artificial neuron to some extent mimics the characteristics of a biological neuron. A set of inputs are presented to the neuron, each weighted by the synaptic value and then summed. The summed value is then passed through an activation function to generate the output of the neuron. The inputs to the neuron are themselves the outputs of other neurons or the input interface to the net, and the output of the neuron will in turn feed on to other neurons or the output interface. Assuming that the network has already learnt, i.e. that is its weights are set to the correct value, then it will be able to recall the learnt exemplar patterns and to some extent generalize when presented with a new set of patterns. Despite the diversity of neural models, nearly all are based on this configuration. Neural models do, however, tend to differ from each other in activation function, network topology and learning algorithm.

The output of a neuron is generally defined by:

$$y = f(\sum_{i=1}^{N} w_i x_i - \theta) \tag{8.1}$$

where w is the synaptic weight to node i, X is the input vector to the node and Q is the bias value. $f(\bullet)$ is the non-linear activation function. Some of the commonly used activation functions are shown in Figure 8.1.

The simplest network is a group of neurons arranged in a layer, as shown in Figure 8.2. The connectivity between the inputs and each neuron may be complete or partial. The single layer of perceptrons is only capable of forming half-plane decision regions. By stacking the single layers to form a multilayer network with non-linear connections used between layers, complex decision regions composed of multiple half-plane regions can be formed.

Application areas for artificial neural network (ANN) models fit into one of the following four classes.

- Pattern classification and associative memory: networks can be trained to distinguish patterns into separate classes and associate input–output pairs.
- Self-organization and feature extraction: using unsupervised learning the network can organize itself to extract features from input data. This is accomplished by adjusting the weights so that topologically close output nodes are sensitive to similar inputs.
- Optimization: the 'energy' of an ANN is defined in terms of its interconnecting weights. By representing a problem in terms of weight intercon-

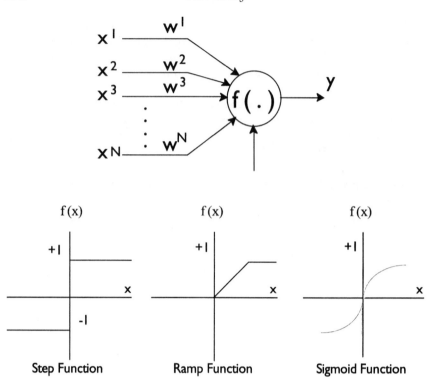

Figure 8.1 Some common activation functions.

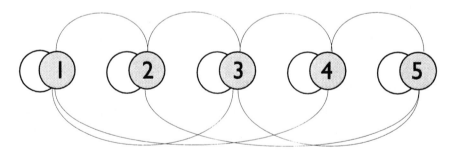

Figure 8.2 Simple network structure.

nections between neurons, an optimized solution to the problem may be obtained by allowing the energy function to converge to a minimum (equilibrium point).

- Non-linear mappings: this is a generalization of the pattern recognition and association properties of ANNs. The ANN is trained so that it learns a non-linear mapping to an input.

8.3 LINEAR ARRAY NEURAL ACCELERATORS

One of the main difficulties with implementing ANNs on digital electronics is the high density of interconnect between neurons, which for a fully interconnected N neuron net scales with $O(N^2)$. In essence, the calculation of a complete neural network function is a repetitive application of a vector–matrix product. The vector elements comprise the current states of the neurons, whereas the matrix holds the weight values associated per row with each neuron synapse. Such operations can be performed efficiently using a linear systolic architecture (a pipeline), which is particularly suited to IC implementation because of its properties of modularity, regularity, local pipelined communications and cascadability. These features have been exploited in several neural accelerator designs. Figure 8.3 shows such a structure.

A linear array consists of a string of simple processing elements each connected to its two nearest neighbours and each provided with its own

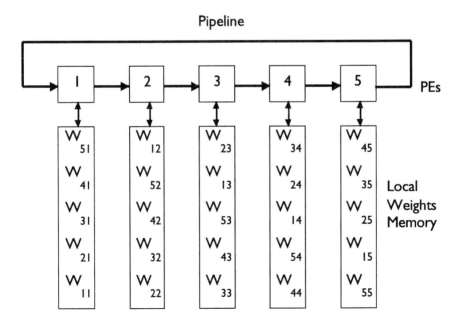

Figure 8.3 Array mapping.

local memory. The memory is used to hold the weight values of connections from other processing elements. The ring array is a particular implementation of the linear array where the two ends of the array are coupled together forming a ring of processing elements (PEs). The basis of the operation is that the outputs of the neurons are piped around the ring so that they visit every other neuron. When the output of one neuron arrives at the input to a second neuron, then the weight representing that connection is loaded in from the second neuron's memory and multiplied by the input. The product is accumulated in the neuron. This occurs for all neurons in the ring so that after one complete rotation all the interconnects have been processed. The number of clock cycles needed to process the interconnect scales with $O(N)$ as opposed to $O(N^2)$ as required for a sequential processor.

8.4 TOROIDAL NEURAL PROCESSOR (TNP)

This section describes the background to the TNP architecture and then proceeds to describe the support it offers for neural model emulation.

8.4.1 Background

The TNP architecture was inspired by the need for a low-cost massively parallel processing system that could emulate a large variety of neural models. The role envisaged for the TNP architecture is to serve as a tool allowing the exploration of neurocomputing problems.

Such an architecture must be flexible enough to satisfy the different forms of interconnect and computational requirements demanded by the various neural models. The suitability of the linear array architecture for providing interconnect flexibility has been addressed above.

8.4.2 Support for neural models

If the TNP design is to support a wide range of neural network models, then it must implement a number of key features. These include:

- provision of a suitable instruction set;
- support for multilayer topologies;
- support for conditional execution of instructions.

The requirement for a simple instruction set was primarily motivated by the desire to observe the primitive operations of neural algorithms. Initial experimentation with various instruction sets indicated that the use of a reduced instruction set numbering around 8–12 instructions can satisfy most of the computational requirements of neural models.

Ideally, the design should implement all the arithmetic operators used for the computation of the neural model set. However, the constraint of a low-cost massively parallel architecture dictates that the design should use a low processor complexity. Hence, only those operators that are most commonly and frequently used, such as multiplication and addition, can be implemented directly. Other less vital operators, such as division and root, are excluded and are instead processed as functions of the current set of operators. The initial selection contains only nine instructions and provides a good basis for the evaluation and selection of a final instruction set. The instruction set is shown in Table 8.1. The inclusion of the random number operator in the instruction set is justified by the need for such functions in stochastic operations such as those used for the Boltzmann and back-propagation networks.

In addition to the PEs, the design of the array controller must also be considered. Being a one-off component it is more cost-effective to use an off-the-shelf microcontroller for the array controller than to develop a custom design. The question that does arise, however, is which microcontroller is most suitable. In order to address this question a basic generic controller design has been assumed and as more advanced capabilities are tested the controller requirements can be refined. Table 8.2 shows the basic instruction set for the assumed array controller.

In keeping with the minimalist design philosophy, all the aforementioned instructions are executed within a single clock cycle, thereby avoiding the need for complex instruction-decoding mechanisms and counters within the PE to keep track of its execution state.

The partitioning of the array into distinct layers can be affected in many ways. One simple yet flexible approach is to use an instruction/PE tagging technique. Each instruction uses a tag field. Instructions are executed within a PE only if the PE tag matches the broadcast instruction's tag. PEs can be clustered together to form layers by assigning them with a common tag. This allows different instructions to be applied to different layers of the network.

Conditionality within the PE is implemented by performing a comparison of the test variable with a reference value. The integer result of the

Table 8.1 TNP processor instruction set

Instruction	Function	Description
NOP	–	No operation
WRITE	MemExt_M[WPtr]	Write MemExt to RAM
READ	M[WPtr]_emExt	Read RAM to MemExt
MUL d,s1,s2	d = s1 x s2	Multiply
ADD d,s1,s2	d = s1 + s2	Add
SUB d,s1,s2	d = s1 - s2	Subtract
TRAN d,s1	d = s1	Transfer s1 to d
RND d,s1	d = RND(s1)	Generate new random number
CMP d,s1,s2	d = Test(s1 3 s2)	d = 1 if s1 3 s2 else 0

Table 8.2 TNP array controller instruction set

Instruction	Function	Description
CNOP	–	No operation
CWRITE	MemExt_S[MPtr]	Write to RAM
CREAD	DS[Mptr]_CMemExt	Read from RAM
CADD d,s1,s2	d = s1 + s2	Add
CSUB d,s1,s2	d = s1 – s2	Subtract
CTRAN d,s1	d = s1	Transfer s1 to d
LOOPNZ	IPtr,Loopn,Countm	Branch on test
IPtr = Loopn iff Count' 0		
STOP	Halt Processor	Send Interrupt to host

comparison operation ('1' representing true, '0' representing false) is placed in an operand register where it is used to select the relevant assignment. An example of a conditional sequence is shown here.

IF X > Y, A= B;
ELSE, A = C;

can be realized using the following operations:

Z = CMP(X,Y); Z = 1 if X > Y, else 0
A = (Z * B) + ((1-Z) * C)   ; calculation complete as Z or (1-Z) = 0

This semi-Boolean expansion of conditional operations assists in achieving deep pipelining of the instruction sequence as no code is skipped and hence the decode–execute pipeline is rarely stalled. Furthermore, this approach reduces the PE complexity and instruction execution time, and maintains the single opcode–operand format.

8.4.3 TNP array

The TNP is designed to be implemented as a rack of cards, each containing many PEs, attached to a workstation bus. Each card will form part of the systolic ring with a separate card containing the array controller, associated memory stores and host interface. The TNP environment is shown in Figure 8.4.

The user wishing to program the TNP can use a graphic interface or a text editor to enter the code either as C source or directly as TNP assembler code. At this current stage in the TNP development, only the assembler entry system is available. The host system preprocesses the assembler code and places the TNP instructions, including both controller and PE instructions, in the controller's instruction store. Input data patterns are placed in a separate data store.

The controller reads instructions from the instruction store and places the output on the instruction bus. If the instruction is a controller instruc-

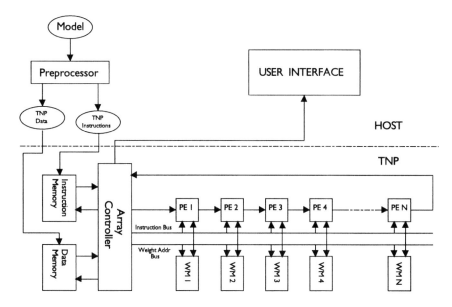

Figure 8.4 TNP environment.

tion then the controller will execute it while the PEs ignore it; conversely the controller will ignore PE instructions and pass them directly onto the PEs.

To keep the PE complexity as simple as possible, the controller handles the weight memory addressing and supplies a common address to all PEs.

Each PE is connected to its two nearest neighbours by a unidirectional data bus that shifts data between PEs from left to right. The output of the last PE is connected to the input of the first PE through the array controller, forming a ring out of the linear array.

8.4.4 Arithmetic precision requirements

The arithmetic functions use a 2's complement 16-bit fixed point representation. The precision of the arithmetic units is particularly crucial. Preliminary experiments with non-trivial problems such as character recognition underlined the need to use 14 bits to specify the fraction portion of the number when using the most widely applied network, back-propagation on multilayer perceptrons. Most other models are satisfied with 8 bits. To accommodate these differences, the TNP PE offers a choice of two radices: 8 bit and 14 bit. For a 16-bit word radix 8 allows a number range of approximately −128.00 to +127.99 (up to two decimal places precision), whereas radix 14 allows a range of around −2.0000 to +1.9999 (four decimal places). Only the 8-bit radix has been implemented in the design in section 8.6.

The fixed-point format is preserved during multiplication by truncating the output. For instance, a 32-bit product is truncated to obtain a 16-bit result by extracting bits 9–24 for radix 8 arithmetic and bits 15–30 for radix 14 arithmetic.

Unfortunately, truncation causes some loss of precision as a result of number overflow. If overflow occurs during truncation the result is saturated, i.e. it is set to the largest positive or negative number, as appropriate. Initial tests with BP also indicated that simply clipping off the lower part of the number during truncation could cause a small weight update value to be incorrectly specified, preventing the network from stabilizing. In order to overcome this problem, rounding is employed. Rounding involves adding a 'I' to the truncated product only if the most significant digit below the truncated portion is 'I'. The need for rounding is also confirmed by several other researchers.

8.4.5 Instruction format

The instruction format for both the TNP PEs and the array controller's instruction word is shown below in Figure 8.5. The word is split up into six separate microcode fields to simplify instruction decoding.

The instruction word size is 20 bits. The C bit indicates whether the instruction is a controller or PE instruction. The Tag field indicates to which PEs the instruction applies (0 = all PEs). 'D' is the destination or result register and SI and S2 are the input operand registers.

8.4.6 Processor architecture

Figure 8.6 shows the structure of the PE. The details of the functional blocks within the PE are described below.

Multiplier

A 16 x 16 bit parallel multiplier has been selected to facilitate high-speed multiplication. Truncation and rounding are employed on the multiplier output. If overflow was detected during truncation, the truncated output is saturated (set to the highest number of the same magnitude that can be

C	Tag	Opcode	D	SI	S2
I	4	3	4	4	4

Figure 8.5 TNP instruction word format.

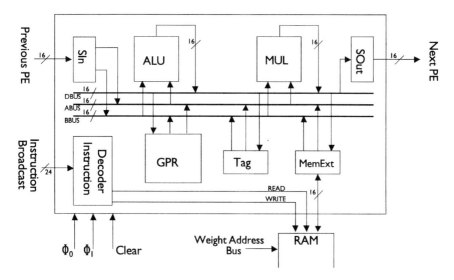

Figure 8.6 TNP processing element.

represented). All operations, including multiplication, overflow detection, truncation, rounding and saturation, occur within the one clock-cycle frame.

ALU (arithmetic logic unit)

A 16-bit adder provides addition and subtraction and comparison operations. Overflow detection and saturation capabilities are also incorporated in the adder. A 16-bit linear feedback shift register process (LFSR) is also employed to provide random numbers.

Communications

Ports SIn and SOut handle communications with nearest neighbours using two **TRAN** instructions. The first **TRAN** instruction is used to route the output from the previous PE in through SIn and the ALU and to store it in the register bank, and the subsequent **TRAN** instruction transfers the data out through the ALU and latches it in SOut, the output port.

Memory

Register **MemExt** is provided to handle data transfers between external RAM and the PE. Control lines **READ** and **WRITE** determine the direction of data flow. The data itself is passed into and out of the register on the MIO bus.

Tag/status register

This register configures the processor operation, determining which tagged instructions it may execute, which number precision it will use (8-bit or 14-bit) and whether rounding is used during multiplication.

GPR (general purpose register)

A bank of nine general-purpose registers (R0–R8) is provided to handle parameter values, state values and temporary work-space. This number was estimated from initial assessments of the benchmark algorithms. One unusual stipulation of the TNP PE design is that a register may not be both source and destination operand for any operation. This simplifies register design and reduces the register count by 1.

Buses

In order to speed up execution, three 16-bit unidirectional buses are used, two of which, ABUS and BBUS, are Read buses connecting any two registers to the Multiplier/ALU, and the third DBUS is the write bus, which places the result in a destination register.

Decoder

The decoding operation is divided into five parallel tasks. The tag field of the broadcast instruction is matched with the contents of the tag register and a flag conditional on the status of the match operation, and the 'C' bit of the instruction word is used to signal whether or not the instruction is valid. The opcode field determines which function is to be executed. The two input operand addresses control the read buses, whereas the destination operand address and the tag-match flag are used to control the write bus.

8.4.7 Array controller

Figure 8.7 shows the chosen structure for the TNP controller with its separate data and instruction stores resembling a Harvard architecture.

Registers CR1–CR8 provide workspace and parameter constant storage. IPtr is the instruction pointer and WPtr holds the address of the current weight location. CSIn and CSOut registers are connected to the first PE's SIn register and last PE's SOut register, respectively, to form the ring. Data from CSIn automatically feeds out to CSOut unless the controller desires to input its own data into the ring. CMemExt is the data transfer register to and from the data store that is addressed via the MPtr register.

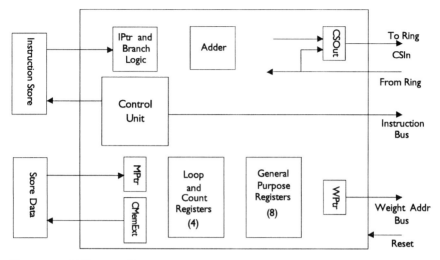

Figure 8.7 TNP controller structure.

The Count and Loop registers are used under control of the LOOPNZ instruction to provide branch conditionality in the instruction stream. The LOOPNZ instruction loads the address from register $Loop_n$ if $Count_m$ is equal to zero. If the condition is not satisfied then IPtr is incremented as usual. The provision of four Count-Loop register pairs allows up to four levels of loop nesting.

8.4.8 Clocking strategy

Figure 8.8 shows the basic clocking strategy. The array is driven by a two-phase non-overlapping clock.

During Φ_0 the controller fetches the next instruction and latches it on the broadcasts bus. A sufficient period is allowed between the falling edge of Φ_0 and the rising edge of Φ for the instruction to settle on the bus and for the PE to execute it.

During Φ_1 the PE latches the data on the Write bus into the destination register. A sufficient period is allowed between the falling edge of Φ_0 and the rising edge of Φ_0 for the instruction to settle on the bus and for the PE to execute it. During Φ_1 the PE also latches the data on the Write bus into the destination register.

8.5 EXAMPLE IMPLEMENTATIONS IN THE TNP

This section demonstrates the complete TNP by showing an example implementation of neural operations.

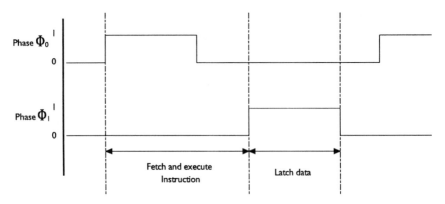

Figure 8.8 Clocking strategy.

Figure 8.9 shows an example of an MLP implementation in a TNP array. The TNP code for forward propagating the code through the hidden layer of this example network is shown in Figure 8.10.

The code presumes that Controller register CR1 contains the correct count to shuffle all the inputs (I) past all the nodes (H) of the hidden layer (H+I – 1). The loop iterates until the branch test is no longer true then breaks out to process the activation function before proceeding to the next layer.

Having introduced the application area and the processor architecture in sufficient detail, the following sections develop a synthesizable design for the TNP processor element

8.6 PROCESSING ELEMENT VHDL DESIGN

The TNP is a simple architecture but consists of many elements. Each element is itself quite complex. This makes it difficult to describe clearly the whole design in a way that will be helpful for the reader. The aim of this design is to:

- demonstrate how a large, multiple element design can be tackled in VHDL; and
- reiterate some of the design complexities already seen in earlier chapters.

The focus of the design will therefore be one element of the TNP – the PE architecture. All the elements of this except two will be considered in the design. First, the multiplier is simply a larger version of the parallel multiplier designed in Chapter 6. It should be clear from this earlier version how to create the required structure and so this is left for the reader to complete. Second, the logic to implement the 8-bit radix has not been constructed (section 8.4.4). This helps to simplify the design of the adder.

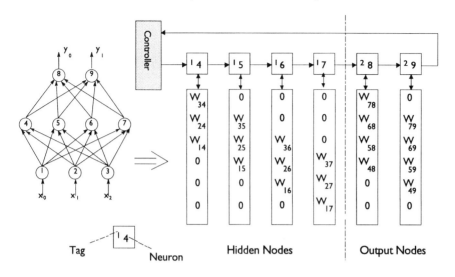

Figure 8.9 Multilayer perceptron (MLP) implementation in the TNP array.

```
TAG 0               ;Address all PEs.
TRAN   R4,R0        ;Clear R4 (R0 = 0).

;Hidden Layer Processing.

CTRAN Count1,CR1    ;CR1 = H+1-1
CTRAN  Loop1,IPtr   ;Return for each iteration.
{
    TAG 0               ;Address all PEs.
    TRAN  R1,SIn        ;Shuffle all PE inputs
    TRAN  SOut,R1       ;one place right.
    TAG 1               ;Address Hidden Layer.
    READ                ;Place contents of Wts memory in MemExt.
    MUL  R2,MemExt,R1       ;R2 = Weight * input.
    ADD   R3,R4,R2          ;Accumulate Products.
    TRAN  R4,R3             ;Update Accumulator.
    CADD  Wptr,Wptr,One     ;Increment Weight pointer.
    CSUB  Count1,Count1,One ;Decrement loop counter.
    LOOPNZ Iptr,Loop1,Count1 ;Repeat again if Count1 ⌐ 0.
}
```

Figure 8.10 Forward-propagation code for the hidden layer of a MLP.

8.6.1 Top-level design

The top-level diagram of the TNP processing element (PE) architecture was shown in Figure 8.6 . This contains each element described in the following sections. Some elements hide lower levels in the hierarchy below these as there are a number of instantiated components. As will be seen, these components reduce the size of the design through replication. In addition to components, some elements contain several processes that segment the design further. Blocks have not been used to generate any extra levels to the hierarchy.

Figure 8.11 contains the top-level schematic of the design. This shows in more detail how the elements of the design are interconnected. A schematic has been used to aid the understanding of the overall design. Each block contains VHDL code only and no further schematics.

The component **REGISTERS** contains all the register and ports that were shown in Figure 8.6. The ALU block has been broken up into its two constituent parts, the adder and the random sequence generator. Also note that the RAM shown in Figure 8.6 has not been designed.

8.6.3 Multiplier

This is instantiated in the top level of the design but the code itself has not been created. The 16 x 16 parallel multiplier is a larger version of the one contained in Chapter 6. Additionally, the multiplier has a rounding capability, overflow and saturation detection. Its design would be a useful exercise for the reader (excluding these features). Be aware, however, of the size of this circuit. One reason that it is omitted here is the length of time the circuit would take to synthesize and the physical size of the schematic that would be produced if the result was to be included in this book.

8.6.3 ALU

The ALU consists of two elements – the adder and the pseudorandom sequence generator. In this design both these elements are present at the top level shown in Figure 8.11. An alternative approach would have been to use components.

Adder

The 16-bit 2's complement adder design is contained in Figure 8.12. This is an extension of the ripple-carry adders designed in Chapter 6. A ripple-carry adder has been used rather than a carry-lookahead adder and so speed has been traded for area. The adder provides addition, subtraction and comparison operations.

The adder itself is a procedure called **RC_ADDER** contained in the package **ARITH_UTILS**. This is accessed as usual through a **Use** clause.

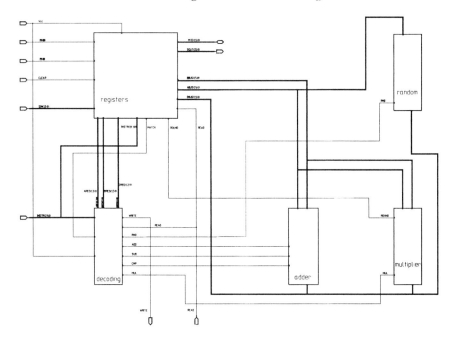

Figure 8.11 Top-level VHDL design for the TNP PE architecture.

The architecture contains a single concurrent call to the **RC_ADDER** procedure. The size of the adder is defined by the size of the array types within the call. **RCOUT** and **OVERFLOW** contain the current sum and carry results for the adder. These values are then used by the irregular **Generate** statement, introduced in Chapter 5, to construct the actual adder result. Three different types of bit are produced – the sign, the radix and the ordinary bits. In each case, the logic is complicated by the overflow detection, saturation and comparison logic. The actual functions are not very important.

The object types in the output assignment statements have been mixed. This has been done purposely in this case to demonstrate the use of the conversion function **TO_X01Z** with the signal **CMP**. Conversion and the similar strength stripper functions were discussed in Chapter 7.

The output of the adder is connected to the internal bus **D**. As the multiplier can also drive this bus, the output must be made three state. Only when the **ADD**, **SUB** or **CMP** instruction signals are active should the result be placed on the bus. The use of three-state logic to handle signals with multiple drivers was introduced in Chapter 7.

The adder itself is contained in Figure 8.13. The **ARITH_UTILS** package contains the two procedures **RC_ADDER** and **FULL_ADDER**. The first defines an N-bit ripple-carry adder and the second a single-bit full adder procedure. These procedures are similar to those constructed as functions

```vhdl
library ARITH, DZX, IEEE;
use ARITH.ARITH_UTILS.all, DZX.LOGIC_UTILS.all,
IEEE.STD_LOGIC_1164.all;

entity ADDER is
   port (
         ADD, SUB, CMP : in BIT;
         ABUS, BBUS : in X01Z_VECTOR(15 downto 0);
         D : out X01Z_VECTOR(15 downto 0)
      );
end ADDER;

architecture STRUCTURAL of ADDER is
   signal RCOUT, RCLOGIC : X01Z_VECTOR(15 downto 0);
   signal OVERFLOW : X01Z;
begin
   RC_ADDER(ABUS, BBUS, SUB, RCOUT, OVERFLOW);
   OUTPUT_LOGIC : for K in 15 downto 0 generate
      -- Account for overflow and comparison operations on each bit
      SIGN_BIT : if (K=15) generate
         RCLOGIC(K)<= (RCOUT(K) xor OVERFLOW) and
TO_X01Z(CMP);
      end generate;
      -- Next bit must be treated differently for comparison operations
      RADIX_BIT : if (K=14) generate
         RCLOGIC(K)<=
            (not(RCOUT(15) or OVERFLOW or TO_X01Z(CMP))) or
            (not OVERFLOW and RCOUT(K) and TO_X01Z(CMP)) or
            (OVERFLOW xor RCOUT(15));
      end generate;
      ORDINARY_BIT : if (K<14) generate
         RCLOGIC(K)<=
            (RCOUT(K) and not OVERFLOW and TO_X01Z(CMP)) or
            (RCOUT(15) and OVERFLOW and TO_X01Z(CMP));
      end generate;
   end generate;

   D <= RCLOGIC when (CMP='0' or SUB='0' or ADD='0') else
      "ZZZZZZZZZZZZZZZZ";

end STRUCTURAL ;
```

Figure 8.12 Top-level logic for the 16-bit adder.

```vhdl
library DZX,IEEE;
use DZX.LOGIC_UTILS.all, IEEE.STD_LOGIC_1164.all;
package ARITH_UTILS is
   procedure FULL_ADDER (A, B, CIN : in BIT; FULLOUT, COUT : out
X01Z);
   procedure RC_ADDER(   A, B : in X01Z_VECTOR; C : in BIT;
                 signal SUM : out X01Z_VECTOR; signal OVER : out X01Z);

end ARITH_UTILS;

package body ARITH_UTILS is

procedure FULL_ADDER (A, B, CIN : in BIT; FULLOUT, COUT : out
X01Z) is
begin
   FULLOUT := TO_X01Z(A xor B xor CIN);
   COUT := TO_X01Z((A or B) and (A or CIN) and (B or CIN));
end FULL_ADDER;

procedure RC_ADDER(   A, B : in X01Z_VECTOR; C : in BIT;
            signal SUM : out X01Z_VECTOR; signal OVER : out X01Z) is
   variable CARRY : X01Z_VECTOR(A'left+1 downto 0);
   variable VSUM : X01Z_VECTOR(SUM'range);
   variable VB : X01Z_VECTOR(B'range);
begin
   VSUM := (others => '0');
   CARRY := (0=>TO_X01Z(C), others=>'0'); -- carry in sets add or
subtract operation
   INVERT : for K in B'range loop
      VB(K) := B(K) xor TO_X01Z(C);   -- invert B for subtraction
   end loop;
   FA : for K in A'reverse_range loop     -- N bit add/subtract operation
      FULL_ADDER (  TO_BIT(A(K)), TO_BIT(VB(K)),
               TO_BIT(CARRY(K)), VSUM(K), CARRY(K+1));
   end loop;

   --check for overflow
   SUM <= VSUM;
   OVER <= CARRY(A'left+1) xor CARRY(A'left);
end RC_ADDER;

end ARITH_UTILS;
```

Figure 8.13 Ripple-carry adder procedures in the ARITH_UTILS package.

in Chapter 6. The key benefit of using the Procedure statement, discussed
in Chapter 7, is that multiple values can be passed back to the calling architecture – both the sum and carry values in this case.

These procedures are very much independent of the TNP design. All the
specialist logic that creates the specific functions required in the ALU is
contained in the ADDER entity and not these procedures. Keeping generic
and specific parts of a design separate in this way is the key to successful
design practices.

Random number generator

A 16-bit linear feedback shift register has been used as a pseudorandom
number generator. The simple function shifts each bit one position to the

```
library IEEE, DZX;
use IEEE.STD_LOGIC_1164.all, DZX.LOGIC_UTILS.all;

entity RAND is
  port (
    ABUS : in X01Z_VECTOR(15 downto 0);
    RND : in X01Z;
    DBUS : out X01Z_VECTOR(15 downto 0) );
end RAND;

architecture BEHAVIOUR of RAND is
    signal REGVAL : X01Z_VECTOR(15 downto 0);
begin
DBUS <= REGVAL;
RAND_GEN :
    process (RND)
        variable SIGNBIT : X01Z;
    begin
      if RND='0' then
          SIGNBIT := (ABUS(0) xor ABUS(2)) xor (ABUS(3) xor
ABUS(5));
          REGVAL <= SIGNBIT & ABUS(15 downto 1);
      elsif RND='1' then
          REGVAL <= REGVAL;
      else
          REGVAL <= (others=>'X');
      end if;
    end process;
end BEHAVIOUR;
```

Figure 8.14 Pseudorandom sequence generator

right and generates a new sign bit. Only if instructed to generate a new random number will the register force a new value on to DBUS. Otherwise the output will be high impedance. The code design for this circuit is shown in Figure 8.14.

8.6.4 Registers and communication ports

The register block contains a variety of general and specific function registers. The top level of this block is the structural design shown in Figure 8.15.

All the required register components are declared in the architecture declarative part of REGISTERS-STRUCTURAL. The actual registers are instantiated in the body of the architecture. The components BASIC_REG and GENERAL_PURPOSE are used for multiple instantiations.

The first is used to create ZERO_REG and SIN_REG (the communications port SIn). The zero register is assigned its value using the Others clause, introduced in Chapter 5. Note that the assignment is not made in the port parameter list itself but instead the signal ZERO is used. The rules for all types of interface lists were discussed in Chapter 5.

The second general register, GENERAL_PURPOSE, is used with a generate statement to create the bank of nine registers. This regular Generate statement taps off decoded instruction bits from AREG, BREG and DREG to control the active register.

The VHDL for each of the register components is shown in Figures 8.16–8.20.

As shown, Basic register is effectively a three-state buffer controlling the values on the A and B buses and not a register. No value is actually stored within the component. The general-purpose register in Figure 8.17 stores a value in GENERALVAL and has a Clear input. The clocked process REG_OP makes use of an incomplete If statement. As explained in Chapter 4, by omitting the Else clause from the 'If DEN='0' then...' statement, the old value of GENERALVAL will be assumed. A feedback loop is therefore inferred in the hardware logic. It is not advisable to leave the logic open to such interpretation, but for this simple design no problems should arise.

The Tag register design in Figure 8.18 is very similar to the general-purpose register. Two extra outputs are derived from this register – MATCH and ROUND. The first goes active to indicate to the PE that a valid instruction has been received. If the incoming tag bits from the instruction, stored in the signal MATCHING, are all zero (indicating a global instruction) or match the value stored in 4 bits of this Tag register the instruction is valid. The ROUND signal simply determines if rounding has to be used during multiplications.

The PR communications port SOut is designed in Figure 8.19. This is a special register with a dual clock and therefore dual clocking processes.

```
library DZX, IEEE;
use DZX.LOGIC_UTILS.all, IEEE.STD_LOGIC_1164.all;

entity REGISTERS is
    port (   AREG, BREG : in X01Z_VECTOR(12 downto 0);
        DREG : in X01Z_VECTOR(11 downto 0);
        READ, PHI0, PHI1, CLEAR : in X01Z;
        TAGBITS : in X01Z_VECTOR(3 downto 0); -- INSTR(18 downto
15)
        DBUS, SIN : in X01Z_VECTOR(15 downto 0);
        ROUND, MATCH : out X01Z;
        ABUS, BBUS, SOUT : out X01Z_VECTOR(15 downto 0);
        MOI : inout X01Z_VECTOR(15 downto 0) );
end REGISTERS;

architecture STRUCTURAL of REGISTERS is

signal ZERO : X01Z_VECTOR(15 downto 0);
-- Instantiate register components
component BASIC_REG
    port (   AEN, BEN : in X01Z;
        OUTVAL : in X01Z_VECTOR(15 downto 0);
        A, B : out X01Z_VECTOR(15 downto 0));
end component;
component MEM_EXT
    port (   AEN, BEN, DEN, READ, CLK, CLEAR : in X01Z;
        D : in X01Z_VECTOR(15 downto 0);
        A, B : out X01Z_VECTOR(15 downto 0);
        M : inout X01Z_VECTOR(15 downto 0) );
end component;
component TAG
    port (   AEN, BEN, DEN, CLK, CLEAR : in X01Z;
        MATCHING : in X01Z_VECTOR(3 downto 0);
        D : in X01Z_VECTOR(15 downto 0);
        ROUND, MATCH : out X01Z;
        A, B : out X01Z_VECTOR(15 downto 0));
end component;
component SOUT
    port (   DEN, CLK0, CLK1, CLEAR : in X01Z;
        D : in X01Z_VECTOR(15 downto 0);
        S : out X01Z_VECTOR(15 downto 0));
end component;
```

Figure 8.15 Top level of the register block.

```
component GENERAL_PURPOSE
   port (    AEN, BEN, DEN, CLK, CLEAR : in X01Z;
      D : in X01Z_VECTOR(15 downto 0);
      A, B : out X01Z_VECTOR(15 downto 0));
end component;

begin
   ZERO <=(others=>'0');
   ZERO_REG : BASIC_REG port map (
               AREG(0), BREG(0),
               ZERO,
               ABUS, BBUS );
   TAG_REG : TAG port map (
               AREG(1), BREG(1), DREG(0),
               PHI0, CLEAR,
               TAGBITS,
               DBUS,
               ROUND, MATCH,
               ABUS, BBUS );
   SIN_REG : BASIC_REG port map (
               AREG(2), BREG(2),
               SIN,
               ABUS, BBUS );
   SOUT_REG : SOUT port map (
               DREG(1),
               PHI0, PHI1, CLEAR,
               DBUS, SOUT );
   MEM_EXT_REG : MEM_EXT port map (
               AREG(3), BREG(3), DREG(2), READ,
               PHI0, CLEAR,
               DBUS,
               ABUS, BBUS,
               MIO );
   GENERAL_REGS : for K in 0 to 8 generate
      RK : GENERAL_PURPOSE port map (
               AREG(K+3),
               BREG(K+3),
               DREG(K+2),
               PHI0, CLEAR,
               DBUS, ABUS, BBUS );
   end generate;

end STRUCTURAL;
```

Figure 8.15 Top level of the register block (continued).

```
library DZX, IEEE;
use DZX.LOGIC_UTILS.all, IEEE.STD_LOGIC_1164.all;
entity BASIC_REG is
   port (   AEN, BEN : in X01Z;
      OUTVAL : in X01Z_VECTOR(15 downto 0);
      A, B : out X01Z_VECTOR(15 downto 0) );
end BASIC_REG;

architecture BEHAVIOUR of BASIC_REG is
begin
   A <= OUTVAL when AEN='0' else "ZZZZZZZZZZZZZZZZ";
   B <= OUTVAL when BEN='0' else "ZZZZZZZZZZZZZZZZ";
end BEHAVIOUR;
```

Figure 8.16 Design of the basic register.

```
library DZX, IEEE;
use DZX.LOGIC_UTILS.all, IEEE.STD_LOGIC_1164.all;
entity GENERAL_PURPOSE is
   port (   AEN, BEN, DEN, CLK, CLEAR : in X01Z;
      D : in X01Z_VECTOR(15 downto 0);
      A, B : out X01Z_VECTOR(15 downto 0) );
end GENERAL_PURPOSE;

architecture BEHAVIOUR of GENERAL_PURPOSE is
   signal GENERALVAL : X01Z_VECTOR(15 downto 0);
begin

   A <= GENERALVAL when AEN='0' else "ZZZZZZZZZZZZZZZZ";
   B <= GENERALVAL when BEN='0' else "ZZZZZZZZZZZZZZZZ";
   REG_OP : process
   begin
      wait until (CLK='1' and CLK'event);
      if CLEAR='0'then
         GENERALVAL <= "0000000000000000";
      else
         if DEN='0' then
            GENERALVAL <= D;
         end if;
      end if;
   end process;

end BEHAVIOUR;
```

Figure 8.17 Design of the general-purpose register.

```vhdl
library DZX, IEEE;
use DZX.LOGIC_UTILS.all, IEEE.STD_LOGIC_1164.all;
entity TAG is
   port (    AEN, BEN, DEN, CLK, CLEAR : in X01Z;
       MATCHING : in X01Z_VECTOR(3 downto 0);
       D : in X01Z_VECTOR(15 downto 0);
       ROUND, MATCH : out X01Z;
       A, B : out X01Z_VECTOR(15 downto 0));
end TAG;
architecture BEHAVIOUR of TAG is
   signal TAGVAL : X01Z_VECTOR(15 downto 0);
begin
   MATCH <=    '1' when (MATCHING="0000" or
MATCHING=TAGVAL(3 downto 0))
         else '0';
   ROUND <= TAGVAL(5);
   A <= TAGVAL when AEN='0' else "ZZZZZZZZZZZZZZZZ";
   B <= TAGVAL when BEN='0' else "ZZZZZZZZZZZZZZZZ";
   REG_OP : process
   begin
      wait until (CLK='1' and CLK'event);
      if CLEAR='0'then
         TAGVAL <= "0000000000000000";
      else
         if DEN='0' then
            TAGVAL <= D;
         end if;
      end if;
   end process;
end BEHAVIOUR;
```

Figure 8.18 Design of the tag register.

This is implemented as a safety measure to prevent rippling of data through the array, which could occur with only a single clock. The first process controls the actual register in the same way as the general registers. The second process uses the second clock to place the register value on the output port of the PE. The important elements of this design are that:

• each clock signal is totally independent of the other, the second is not derived from the first; and
• the signals assigned in each process are unique to that process.

The rules for the creation of synchronous sections with both Wait and If statements were fully explained in Chapter 5. The Dos and Don'ts section

```
library DZX, IEEE;
use DZX.LOGIC_UTILS.all, IEEE.STD_LOGIC_1164.all;

entity SOUT is
   port (    DEN, CLK0, CLK1, CLEAR : in X01Z;
          D : in X01Z_VECTOR(15 downto 0);
          S : out X01Z_VECTOR(15 downto 0)
          );
end SOUT;
architecture BEHAVIOUR of SOUT is
   signal SOUTVAL : X01Z_VECTOR(15 downto 0);
begin

   REG_OP1 : process
   begin
      wait until (CLK0='1' and CLK0'event);
      if CLEAR='0'then
        SOUTVAL <= "0000000000000000";
      else
        if DEN='0' then
           SOUTVAL <= D;
        end if;
      end if;
   end process;

   -- control output with second clock to filter array skew/glitches
   REG_OP2 : process
   begin
      wait until (CLK1='1' and CLK1'event);
      if CLEAR='0'then
        S <= "0000000000000000";
      else
        S <= SOUTVAL;
      end if;
   end process;
end BEHAVIOUR;
```

Figure 8.19 Design of the SOut port.

of that chapter also covered some fundamental issues relating to signal assignments within processes. The reader is encourage to review these.

The final register, MEM_EXT, interfaces to external RAM. Its VHDL code is shown in Figure 8.20. It is essentially the same as a general-purpose register except that the input will be derived from the D bus if DREG(2) is low

```
library DZX, IEEE;
use DZX.LOGIC_UTILS.all, IEEE.STD_LOGIC_1164.all;
entity MEM_EXT is
   port (    AEN, BEN, DEN, READ, CLK, CLEAR : in X01Z;
       D : in X01Z_VECTOR(15 downto 0);
       A, B : out X01Z_VECTOR(15 downto 0);
       M : inout X01Z_VECTOR(15 downto 0) );
end MEM_EXT;
architecture BEHAVIOUR of MEM_EXT is
   signal MEXTVAL : X01Z_VECTOR(15 downto 0);
begin
   M <= MEXTVAL;
   A <= MEXTVAL when AEN='0' else "ZZZZZZZZZZZZZZZZ";
   B <= MEXTVAL when BEN='0' else "ZZZZZZZZZZZZZZZZ";
   REG_OP : process
   begin
      wait until (CLK='1' and CLK'event);
      if CLEAR='0'then
        MEXTVAL <= "0000000000000000";
      else
        if DEN='0' then
           MEXTVAL <= D;
        elsif READ='0' then
           MEXTVAL <= M;
        end if;
      end if;
   end process;
end BEHAVIOUR;
```

Figure 8.20 Design of the Memext register.

or from the external memory on the M bus if READ is low. Again the incomplete If statement has been used to infer that if both control signals are high the present value of MEXTVAL is retained.

8.6.5 Instruction decoder

Decoders were designed in both Chapters 4 and 7. Those required to decode the instructions for the TNP PE are very similar to these.

The main architecture of the decoder is shown in Figure 8.21. CONTROL-BEHAVIOUR only processes the incoming 20-bit instruction when the MATCH and C (INSTR(19)) bits indicate that it is destined for this PE. The first action is to enable the function required – read, write, multiply and so

```
library DZX, IEEE;
use IEEE.STD_LOGIC_1164.all, DZX.LOGIC_UTILS.all;
entity CONTROL is
   port (   INSTR : in X01Z_VECTOR(19 downto 0); MATCH : in X01Z;
      MUL, CMP, SUB, ADD, RND, READ, WRITE : out X01Z;
      BREG, AREG : out X01Z_VECTOR(13 downto 0);
      DREG : out X01Z_VECTOR(12 downto 0));
end CONTROL;
architecture BEHAVIOUR of CONTROL is
   component DECODER4_16
      port( signal I : in X01Z_VECTOR(3 downto 0);
         signal Y : out X01Z_VECTOR(15 downto 0));
   end component;
   signal DEST_ENABLE, INST_ENABLE, WRITE1, READ1, NOP,
      A1, A2, A3, B1, B2, B3 : X01Z;
   signal Y_MAP_A, Y_MAP_B : X01Z_VECTOR(15 downto 0);
begin
INSTRUCTION : process      -- decode the actual instruction
   begin
      wait until (INST_ENABLE='0');
         WRITE1 <= '1';
         READ1 <= '1';
         NOP <= '1';
         MUL <= '1';
         ADD <= '1';
         SUB <= '1';
         RND <= '1';
         CMP <= '1';
         case INSTR(14 downto 12) is
            when "000" => NOP <= '0';
            when "001" => WRITE1 <= '0';
            when "010" => READ1 <= '0';
            when "011" => MUL <= '0';
            when "100" => ADD <= '0';
            when "101" => SUB <= '0';
            when "110" => RND <= '0';
            when "111" => CMP <= '0';
            when others =>
               NOP <= 'X';
               WRITE1 <= 'X';
               READ1 <= 'X';
               MUL <= 'X';
               ADD <= 'X';
               SUB <= 'X';
```

Figure 8.21 Design of the instruction decoder.

```
DESTINATION : process
  begin
    wait until (DEST_ENABLE='0');
      case INSTR(11 downto 8) is
        -- no 0000, 0010 or 1111 combinations
        when "0001" => DREG <= "1111111111111110";
        when "0011" => DREG <= "1111111111111101";
        when "0100" => DREG <= "1111111111111011";
        when "0101" => DREG <= "1111111111110111";
        when "0110" => DREG <= "1111111111101111";
        when "0111" => DREG <= "1111111110111111";
        when "1000" => DREG <= "1111111101111111";
        when "1001" => DREG <= "1111110111111111";
        when "1010" => DREG <= "1111101111111111";
        when "1011" => DREG <= "1111011111111111";
        when "1100" => DREG <= "1101111111111111";
        when "1101" => DREG <= "1011111111111111";
        when "1110" => DREG <= "0111111111111111";
        when others => DREG <= "1111111111111111";
      end case;
  end process;
-- internal signal logic
INST_ENABLE <= not INSTR(19) nand MATCH;
DEST_ENABLE <= (NOP nand WRITE1 nand READ1) nor (not
INSTR(19) nand MATCH);

-- output signal logic and mapping

SOURCE_A : DECODER4_16 port map (INSTR(7 downto 4), Y_MAP_A);
SOURCE_B : DECODER4_16 port map (INSTR(3 downto 0), Y_MAP_B);
AREG(0) <= Y_MAP_A(15) and Y_MAP_A(3) and Y_MAP_A(0);
BREG(0) <= Y_MAP_B(15) and Y_MAP_B(3) and Y_MAP_B(0);
AREG(13 downto 3) <= Y_MAP_A(14 downto 4);
BREG(13 downto 3) <= Y_MAP_B(14 downto 4);
AREG(2 downto 1) <= Y_MAP_A(2 downto 1);
BREG(2 downto 1) <= Y_MAP_B(2 downto 1);
WRITE <= WRITE1;
READ <= READ1;

end BEHAVIOUR;
```

Figure 8.21 Design of the instruction decoder (continued).

on. The **Case** instruction must cover all possible permutations of the expression being evaluated. When it was first introduced in Chapter 4, only objects of type **BIT** were used. In Chapter 7, a complication was introduced when new multivalued types were discussed. The **Others** clause is necessary in this **Case** statement to cover all the additional permutations that the object type **X01Z_VECTOR** contains. The reality for the hardware implementation is, however, that this clause will be ignored. Only during the simulation will it have any effect on the output of the decoder. In this instance, a 'don't care' output will indicate an invalid instruction bit. This design will exhibit different simulation characteristics before and after synthesis because of this statement. The Dos and Don'ts section of Chapter 7 discusses this point in more detail.

A second process in the instruction decoder is the destination decoder. This selects which register will be the destination of the value about to be calculated or generated. By referring to the register block code in Figure 8.15, the mapping of **DREG** bits into the registers can be seen. The **Zero** and **Sin** registers cannot be written to, so not all combinations of the four instruction bits must be decoded. These are handled through an **Others** clause that forces all bits high.

The mapping of source registers into **AREG** and **BREG** can also be seen from Figure 8.15. The decoding of the appropriate instruction bits is identical for both and so a 4- to 16-line decoder component is used. This is **DECODE4_16**, shown in Figure 8.22. The output from this cannot be used directly but must be shifted and/or mapped again for some bits. This could have been done within the decode component, but it would then have lost its generic form. The extra mapping is required simply because three instructions all require the **Zero** register as an input.

8.7 LOGIC SYNTHESIS AND OPTIMIZATION

The approach taken in previous chapters of presenting the synthesized and optimized design results has not been used here. The TNP design is larger and more complex than examples in earlier chapters. The key reason for including synthesis and optimization results has been to demonstrate how VHDL constructs are mapped into hardware and how optimization can change the hardware that is generated. The constructs used in this case study are examples of many already seen. This section will suggest the structures that will result from synthesis but will not present the actual results. Readers are left to attempt the synthesis themselves and are referred to earlier examples for comparisons.

8.7.1 Top-level design

The top level is a structural design. It contains the four components that make up the PE. Instantiating components at this high level is a common

```vhdl
library DZX, IEEE;
use IEEE.STD_LOGIC_1164.all, DZX.LOGIC_UTILS.all;

entity DECODER4_16 is
   port (
     I : in X01Z_VECTOR (3 downto 0);
     Y : out X01Z_VECTOR (15 downto 0) );
end DECODER4_16;

architecture DATAFLOW of DECODER4_16 is
begin
   with I select
   Y <= "1111111111111110" when "0000",
        "1111111111111101" when "0001",
        "1111111111111011" when "0010",
        "1111111111110111" when "0011",
        "1111111111101111" when "0100",
        "1111111111011111" when "0101",
        "1111111110111111" when "0110",
        "1111111101111111" when "0111",
        "1111111011111111" when "1000",
        "1111110111111111" when "1001",
        "1111101111111111" when "1010",
        "1111011111111111" when "1011",
        "1110111111111111" when "1100",
        "1101111111111111" when "1101",
        "1011111111111111" when "1110",
        "0111111111111111" when "1111",
        "1111111111111111" when others;

end DATAFLOW;
```

Figure 8.22 Design of the 4- to 16-line decoder.

way of combining the elements of a design. The structural designs of the 4 to 1 multiplexer in Chapter 4 did exactly that, even though the components themselves were small. The important difference here is that each block is unique. Flattening the structure would probably reduce the circuit area a small amount, but preserving it is vital to the understanding of the function. Furthermore, it is important to optimize different elements in different ways. The dataflow logic must not be treated in the same way as control logic.

8.7.2 ALU

The arithmetic logic unit consists of the adder and the pseudorandom sequence generator. These are two independent units that are combined at the second level of the hierarchy through component instantiations. Internally, each is very different.

The generator is a block of sequential control logic contained within a process. The key requirement for its synthesis and optimization is that area is minimized. Timing constraints can then be applied if required.

The adder is one of the main dataflow elements in the design. The multiplier is the other. Unlike the control logic its structure is important – it is designed as a ripple-carry adder for a reason! A variety of adders were designed in Chapter 6 and a great emphasis was placed on maintaining the circuits' physical characteristics after optimization. The structure of the carry-lookahead adder was particularly complex and it would be interesting for the reader to review the way in which the circuit was synthesized and optimized. The timing characteristics are obviously the most important for any data-driven circuit.

This adder is 'constructed' with procedure calls. The synthesized design will therefore not contain any blocks that would help to segregate the full-adder elements and the output control logic. A simple **Block** statement (discussed in Chapter 7) placed around the output logic would help this situation. To complicate the situation further, a procedure is compiled but not synthesized before it is used. Therefore, unlike a component, it cannot be individually optimized before being incorporated in the design. Again Chapter 6 should provide guidance on the problems that this circuit presents to the optimizer (and the designer!).

8.7.3 Registers and communications ports

The simplest of components are the ones that benefit the least from the power of the optimizer. The register block is no exception.

Registers are used because of the speed of access that they offer compared with conventional memory, but the penalty is the area that each one occupies. In this design, this problem is complicated by the use of a number of special and unique registers that must be individually optimized. In every case, minimizing the area is important, but ensuring that the performance benefits are realized by optimizing for speed is equally vital.

The top level of the register block should be synthesized so that its clear structure is maintained. The registers themselves will all contain sub-blocks because of the **Process** statements each uses. Area optimization will remove these, but as **SOut** requires two independent clocks its two module structure should ideally be retained.

The clocked processes infer storage elements. The counters designed in Chapter 5 investigated the types of D-type flip flops that are implemented under different circumstances. It should be easy to predict which will be used in these registers.

8.7.4 Instruction decoder

The instruction decoder is similar to the type of logic seen in both Chapters 4 and 7. The combinational logic that it contains will create a large circuit when synthesized but should be easy to optimize. However, this will rely on the structured design that the components and processes create by being flattened. Only in this way can the optimizer minimize the duplication across the whole logic block.

Judging whether the circuit should be optimized as a complete block or as individual components is often hard. Although the total area of a flattened circuit may be smaller, the critical path may become complex and difficult to manage. If low, medium and then high levels of optimization effort are applied the path may interconnect between several originally independent control functions and the relationships between component paths and fan-in or -out can become hard to understand.

8.8 SUMMARY

This chapter has been a case study of the processing element architecture of the toroidal neural processor. The simple aim has been to present a complete design that incorporates a wide range of the VHDL constructs and synthesis ideas and issues that were introduced in the earlier chapters. It is hoped that the reader will perform some of the suggested exercises linked to this case study and discussed in the text.

9

Summary and future directions

9.1 INTRODUCTION

This chapter aims to detail the key points brought out in this book for the design of effective and efficient VHDL designs. At each of these points, the reader is referred back to the appropriate section where more detailed exposition of the points can be found. This chapter can therefore be used as a short guide to the key messages to be found in this book. Following this, the final section identifies some trends likely to occur in the development of synthesis tools as a result of the technology push associated with the move to deep submicron technology and the user push associated with improving designer productivity.

9.2 THE VHDL APPROACH TO ELECTRONIC SYSTEMS DESIGN

VHDL is a way of describing the operation of an electronic system. As such, VHDL permits the description of a system in a number of different ways. Furthermore, each of these different ways of description (behavioural, dataflow and structural) can be mixed within the same design. The particular use of the different approaches is not a straightforward choice to make. Furthermore, it is rare to find within even a moderately complex design the use of only one design style. Generally speaking, for designs in which circuit complexity and speed are not considerations, the behavioural approach offers the benefit of improved productivity gains. The dataflow and structural approaches need to be used both when specific design constructs are being developed (e.g. a state machine) and where speed and/or performance is important. However, as we have said, the exact choice of approach is not straightforward and relies to some substantial extent on the designer's experience and judgement. In practice, the VHDL designer must be able to use all three approaches with equal fluency.

As we have said, VHDL is a language that enables both the function and the interconnect (when appropriate) of a system to be described. VHDL is a highly structured language with very strong typing. In particular, it is important to understand the difference in concept between signals and variables. Most importantly, the ability to understand when and where wires and storage elements are instantiated is essential (sections 5.1 and 5.4).

9.2.1 Sequential and concurrent constructs

Sequential constructs enable the designer to describe the behaviour of the systems without specifying exactly the logic and interconnect required to achieve it. This is in contrast to the concurrent constructs, which go further down the path of describing discrete logic elements or modules with a clearer (if not fully defined) understanding of the physical circuit itself.

Chapters 4 and 5 show that the design of sequential and combinational logic is related at least in part to the sequential and parallel constructs respectively. By this we mean that when designing combinational logic it is easier (we would argue) to understand and describe such systems using parallel VHDL statements. In addition, we would argue that parallel VHDL constructs simplify the design of complex and/or regular arithmetic circuits due to the ability to express iterative designs using 'loops' (section 6.3.2).

Sequential constructs introduce many issues and concepts before a complete understanding is possible. Key ones include the following.

- How to create clocking statements (section 5.2.2).
- When and where to use clock statements. Especially important here is the understanding of the likely structure needed before designing the circuit logic. This will determine whether it is better to instantiate or infer storage elements (sections 5.1.2 and 5.1.3).
- Restrictions on the content of clock statements. It is far too easy to create storage elements that require components which may it may not be possible to synthesize (e.g. a device with both a synchronous clear and asynchronous preset) using a particular library. (Chapter 5 showed how a number of different D-types could be inferred.)
- Understanding the use of signals and variables within sequential logic blocks and where and when values are updated (section 5.42).
- The different triggering mechanisms for storage elements (e.g. rising, falling and level triggered devices) and the importance of having a consistent timing philosophy to enable the smooth hazard-free transfer of data within a circuit. Chapter 5 demonstrated a number of clocking schemes and the state machines of Chapter 7 illustrated the importance of carefully designed clock structures.

9.2.2 Building block approach

Many design disciplines promote the use of modular design and building blocks to enhance design productivity, and VHDL is no exception. The building block approach simplifies the design process by splitting up the task into more easily comprehensible pieces. The reuse of modules constructed by other projects clearly improves design productivity.

The particular VHDL constructs that support this include: component instantiations, where other design elements can be seamlessly incorporated into a design (section 4.2.2); the use of procedures and functions to both partition the design task and facilitate top-down design techniques (section **6.2.2** for functions and section 7.3.2 for procedures); the use of blocks to structure the design description (section 7.2.2).

Technology vendors such as the semiconductor manufacturers produce these flexible libraries based on building blocks for their own products in order to make the design task simpler and the synthesis more efficient.

9.2.3 Objects and types

Signals, variables and constants are at the heart of every design. They were introduced in Chapter 3. There are significant differences between these objects in terms of the way in which they can be used to hold and pass data. Chapter 5 demonstrated these differences with particular emphasis on signals and variables.

Objects must also have a type. This determines how they can interact not only with different objects but also with similar objects such as two signals. Two rules for making assignments exemplify the limitations.

- A value can only be assigned to an object if it is contained within the definition of the object's type.
- An object can only be assigned to another object if it is of the same type or is a subtype of that type.

The use of different types, both scalar and composite, was introduced in Chapter 3. As the second rule above mentions however, objects may be defined not as a type, but rather as a subtype. This further restricts how the object can be used (section 5.4.1). The benefit of this approach is that the designer can call upon the library functions that are available for the type, thereby saving significant effort.

9.3 SYNTHESIS USING VHDL

The key message of this book is design for synthesis. Examples throughout the book have illustrated not only **which** synthesized circuits are generated but also detailed why a circuit designed for simulation will not necessarily synthesize (e.g. the Dos and Don'ts of Chapters 4–7 and in particular section 7.1).

A number of VHDL language constructs have not been discussed (e.g. do something **AFTER 3 nS**). Similarly, a number of the constructs that have been covered are restricted in how they can be used. (The **Block** statement in Chapter 7 and the **Wait** statement in Chapter 4 are key examples.)

Applying the language to designing circuits that may be synthesized is not difficult, but the following key points should be remembered.

- Design for synthesis – this has been the ethos of the book. Using VHDL for synthesis is where its greatest potential lies. When using constructs for the design of sequential and concurrent logic, often the user has to have a previously conceived outline of the design solution. Except for the simplest designs, the capability of contemporary synthesis tools to create a design without any guidance as to structure is limited. Hence, the designer needs to have a loose design structure in mind before coding.
- VHDL is sequential, concurrent or both. Depending on where you place statements such as signal assignments, different logic can result. Remember that just because an assignment statement appears above another one you cannot assume that the assignment has actually occurred. The rules for sequential constructs are very specific on the use of signal assignments. Using variables is a safer and clearer way of constructing such statements, as the examples have demonstrated (Chapters 5 and 7).
- Concurrent statements require similar careful thought. Multiple-signal assignments in concurrent statements will create multiple signal drivers. Three-state drivers or wired logic are required to implement multiple signal assignments (section 7.2.3).
- Identical functionality does not mean identical circuits. A behavioural VHDL description that uses a Case statement can have an identical function to a description constructed from 'if-then' statements. However, as the synthesized circuits illustrated in Chapter 4, the characteristics of the circuits generated are very different, although functionally identical. Chapter 4 further illustrated that the same behaviour can be created using either concurrent or sequential logic constructs. The choice of statement type can have a significant impact on circuit characteristics.

9.4 FUTURE DEVELOPMENTS

9.4.1 Technology

The microelectronics industry is a fast-moving industry and a number of developments are likely to take place in the next 5 years. Technology will continue to improve. Device feature sizes will continue to shrink. In itself this will continue to bring the increased packing density per chip, the improvement in clock speed and the improvement in power consumed per transistor. Concurrently, the problems of testability of such tightly integrated systems and the overall power consumption per device will continue to pose problems. However, in addition to these problems, a new one will be faced that may pose substantially new problems for the designers of synthesis tools.

Figure 9.1 shows a graph relating gate delay to interconnection delay for a range of processes.

Down to 1 μm feature size, the delays in the circuit are primarily caused by the switching speed of the gate and the capacitance associated with the gates. However, as the transistors shrink in size, their switching speed improves, but the resistance of the interconnect gets relatively worse. As feature sizes move towards 0.1 μμ, the majority of the delay in a system is due to interconnect delay.

For synthesis tools, the above developments turn the principles of effective and efficient synthesis upside down. Previously, synthesis tools concentrated on the design of small and fast blocks of logic, which are connected together via wires. In contrast, future synthesis tools are likely to need to construct small and fast blocks of interconnection and attach blocks of logic to these networks. Conceptually, this introduces the issues of physical location as a design parameter, whereas previously synthesis was concerned almost primarily with logic minimization and critical path reduction.

Synthesis tools for deep submicron design (less than 0.3 μμ) will have to deal much more with the issue of timing. They will need to partition designs and allocate design resources (in this case wires as much as gates) on the basis of reducing physical distance between components.

These developments will mean that synthesis tools may have to work closely with floorplanning tools, i.e the designer of the system will have to have at least a high-level idea of the physical mapping of the system. Of course, CAD tools will support the designer in the partitioning process (as current synthesis tools support the designer in the construction of the system logic), but, nonetheless, the domination of delay time by interconnect length will force the designer and the design tools to address physical layout.

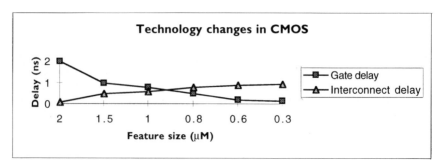

Figure 9.1 Gate and interconnect delay for a range of processes.

It was briefly alluded to in the introduction of this section that the power consumed per chip will increase as a result of the greater number of devices per chip. This fact in itself (the increased power density of ICs) will also affect the design of synthesis tools.

It is becoming increasingly important to reduce the power consumption of advanced integrated circuits. There are four main causes of power consumption in a CMOS circuit: firstly, the short-circuit current caused when a gate switches; secondly, the power consumption caused by charging the capacitances associated with each gate; thirdly, the capacitance of the interconnect between gates; and, fourthly, the capacitance associated with the clock lines. In deep submicron design the third and fourth factors will appear to dominate. Clearly, the use of floorplanning techniques for performance will also reduce this factor. However, there is another issue to consider, which is the synthesis of the logic with a view to minimizing power consumption. For example, in an adder circuit, each node may change state more than once during a single addition as a result of the ripple of the carry circuit passing along each bit stage. At each transition energy is expended. It may well be important in future deep submicron designs to synthesize logic to minimize the power consumption of the system.

9.4.2 Hardware description languages

VHDL is an international standard. There is much to be gained from the standardization of the language, almost independent of any merits of the language itself. In this sense, VHDL is likely to continue as a design standard for very many years to come. There will certainly be amendments and upgrading over the years, as has happened to many international computer languages (e.g. FORTRAN). At the same time, the benefits of language-based design will be applied to the analogue design domain, probably via the language AHDL (analogue hardware description language). At the same time, synthesis techniques are likely to improve, enabling more complex constructions to be synthesized automatically and higher level design entry to become possible. Another important development is the VITAL initiative developing standards for common library definitions

Perhaps the greatest developments will come from the use of HDLs for the entire design process. Hardware–software co-design is the process whereby both the electronic circuitry and the software of a circuit are developed simultaneously. Clearly, more and more designs incorporate both hardware and software elements. The push for reducing design time and increasing capability is causing designers to address the use of language-based design. The synthesis tools addressed in this book have dealt only with producing hardware. Emerging out of research laboratories and into the marketplace are co-design tools that can analyse and partition a design into a good mixture of software components and hardware components. We await their wide use and the development of books teaching the use of these tools with great interest.

9.5 CONCLUDING REMARKS

This book has tried to develop and teach a range of techniques for the construction of complex electronic systems using VHDL and logic synthesis tools. The essential principles are clear – simple language constructs allied to an understanding of likely hardware constructs. The strength and beauty of good design is often to be found in its simplicity not its complexity. We wish you much simplicity in your designs!

Appendix A
Predefined language environment

RESERVED WORDS

The identifiers listed below are called **reserved words** and are reserved for significance in the language. These words cannot be used as explicit identifiers. VHDL is case insensitive.

abs	exit	new	severity
access		next	shared
after	file	nor	signal
alias	for	not	sla
all	function	null	sll
and			sra
architecture	generate	of	srl
array	generic	on	subtype
assert	group	open	
attribute	guarded	or	then
		others	to
begin	if	out	transport
block	impure		type
body	in	package	
buffer	inertial	port	unaffected
bus	inout	postponed	units
	is	procedure	until
case		process	use
component	label	pure	
configuration	library		variable
constant	linkage	range	
	literal	record	wait
disconnect	loop	register	when
downto		reject	while
	map	rem	with
else	mod	report	
elsif	nand	return	xnor
end		rol	xor
entity		ror	
		select	

PREDEFINED OPERATORS

The operators that may be used in expressions are defined below. Each operator belongs to a class of operators, all of which have the same precedence level; the classes of operators are listed in order of increasing precedence. The precedence of operators is fixed.

logical_operator	and	or	nand	nor	xor	xnor
relational_operator	=	/=	<	<=	>	>=
shift_operator	sll	srl	sla	sra	rol	ror
adding_operator	+	-	&			
sign	+	-				
multiplying_operator	*	/	mod	rem		
miscellaneous_operator		**	abs	not		

PREDEFINED ATTRIBUTES

The predefined attributes that can be used for synthesis are given below. Other attributes exist but cannot be used during synthesis. This list has been compiled from the *IEEE Standard VHDL Language Reference Manual*.

Name (kind)	Prefix	Parameters	Result
T' LEFT (value)	Any scalar type or subtype T		The left bound of T. Same type as T
T' RIGHT (value)	Any scalar type or subtype T		The right bound of T. Same type as T
T' HIGH (value)	Any scalar type or subtype T		The upper bound of T. Same type as T
T' LOW (value)	Any scalar type or subtype T		The lower bound of T. Same type as T
A' LEFT[(N)] (function)	Any prefix A that is appropriate for an array object, or an alias thereof, or that denotes a constrained array subtype	A locally static expression of type universal_integer, the value of which must not exceed the dimensionality of A. If omitted, it defaults to 1.	Left bound of the Nth index range of A. (If A is an alias for an array object, then the result is the left bound of the Nth index range from the declaration of A, not that of the object). The result type is the type of the left bound of the Nth index range of A.
A' RIGHT[(N)] (function)	Any prefix A that is appropriate for an	A locally static expression of type	Right bound of the Nth index range of A. (If A

	array object, or an alias thereof, or that denotes a constrained array subtype	universal_integer, the value of which must not exceed the dimensionality of A. If omitted, it defaults to I.	is an alias for an array object, then the result is the right bound of the Nth index range from the declaration of A, not that of the object). The result type is the type of the right bound of the Nth index range of A.
A' HIGH[(N)] (function)	Any prefix A that is appropriate for an array object, or an alias thereof, or that denotes a constrained array subtype	A locally static expression of type universal_integer, the value of which must not exceed the dimensionality of A. If omitted, it defaults to I	Upper bound of the Nth index range of A. (If A is an alias for an array object, then the result is the upper bound of the Nth index range from the declaration of A, not that of the object). The result type is the type of the Nth index range of A.
A' LOW[(N)] (unction)	Any prefix A that is appropriate for an array object, or an alias thereof, or that denotes a constrained array subtype	A locally static expression of type universal_integer, the value of which must not exceed the dimensionality of A. If omitted, it defaults to I	Lower bound of the Nth index range of A. (If a is an alias for an array object, then the result is the lower bound of the Nth index range from the declaration of A, not that of the object). The result type is the type of the Nth index range of A.
A' RANGE[(N)] Range	Any prefix A that is appropriate for an array object, or an alias thereof, or that denotes a constrained array subtype	A locally static expression of type universal_integer, the value of which must not exceed the dimensionality of A. If omitted, it defaults to I	The range A' LEFT(N) to A' RIGHT(N) if the Nth index range of A is ascending, or the range A' LEFT(N) downto A' RIGHT(N) if the Nth index range of A is descending. (If A is an alias for an array object, then the result is determined by the Nth index range from the declaration of A, not that of the object). The result type is the type of the Nth index range of A.

A' REVERSE RANGE (range)	Any prefix A that is appropriate for an array object, or an alias thereof, or that denotes a constrained array subtype	A locally static expression of type universal_integer, the value of which must not exceed the dimensionality of A. If omitted, it defaults to I	The range A' RIGHT(N) downto A' LEFT(N) if the Nth index range of A is ascending, or the range A' RIGHT(N) to A' LEFT(N)) if the Nth index range of A is descending. (If A is an alias for an array object, then the result is determined by the Nth index range from the declaration of A, not that of the object). The result type is the type of the Nth index range of A.
A' LENGTH (value)	Any prefix A that is appropriate for an array object, or an alias thereof, or that denotes a constrained array subtype	A locally static expression of type universal_integer, the value of which must not exceed the dimensionality of A. If omitted, it defaults to I	Number of values in the Nth index range i.e. if the Nth index range of A is a null range, then the result is 0. Otherwise, the result is the value of T' POS(A' HIGH(N)) − T' POS (A' LOW(N)) +I, where T is the subtype of the Nth index of A. The result type is unversal_integer.
S' STABLE (signal))	Any signal denoted by the static signal name S	A static expression of type TIME that evaluates to a non-negative value. If omitted, if defaults to 0 ns	A signal that has the value TRUE when an event has not occurred on signal S for T units of time and the value FALSE otherwise. The result type is Boolean.
S' EVENT (function)	Any signal denoted by the static signal name S		A value that indicates whether an event has just occurred on signal S. The result type is Boolean.
S' LAST_VALUE (function)	Any signal denoted by the static signal name S		The previous value of S, immediately before the last change of S. The result type is the base type of S.

Appendix B

Predefined packages and libraries

STANDARD PACKAGE

The STANDARD package predefines a number of type, subtypes and functions. The package is implicitly included in each design file. The package cannot be modified.

All the definitions contained within the package are shown below. However, not all can be used when designing for synthesis. Refer to relevant chapters for more details.

```
package STANDARD is
    type BOOLEAN is (FALSE, TRUE);
    type BIT is ('0','1');
    type CHARACTER is (
        NUL,  SOH,  STX,  ETX,  EOT,  ENQ,  ACK,  BEL,
        BS,   HT,   LF,   VT,   FF,   CR,   SO,   SI,
        DLE,  DC1,  DC2,  DC3,  DC4,  NAK,  SYN,  ETB,
        CAN,  EM,   SUB,  ESC,  FSP,  GSP,  RSP,  USP,
        ' ',  '!',  '"',  '#',  '$',  '%',  '&',  ''',
        '(',  ')',  '*',  '+',  ',',  '-',  '.',  '/',
        '0',  '1',  '2',  '3',  '4',  '5',  '6',  '7',
        '8',  '9',  ':',  ';',  '<',  '=',  '>',  '?',
        '@',  'A',  'B',  'C',  'D',  'E',  'F',  'G',
        'H',  'I',  'J',  'K',  'L',  'M',  'N',  'O',
        'P',  'Q',  'R',  'S',  'T',  'U',  'V',  'W',
        'X',  'Y',  'Z',  '[',  '\',  ']',  '^',  '_',
        '`',  'a',  'b',  'c',  'd',  'e',  'f',  'g',
        'h',  'i',  'j',  'k',  'l',  'm',  'n',  'o',
        'p',  'q',  'r',  's',  't',  'u',  'v',  'w',
        'x',  'y',  'z',  '{',  '|',  '}',  '~',  DEL );
    type SEVERITY_LEVEL is (NOTE, WARNING, ERROR, FAILURE);
```

-- continued

-- STANDARD package continued

```
type INTEGER is range implementation_defined;
type REAL is range implementation_defined;
type TIME is range implementation_defined
   units
      fs;
      ps   =    1000 fs;
      ns   =    1000 ps;
      us   =    1000 ns;
      ms   =    1000 us;
      sec  =    1000 ms;
      min  =    60 sec;
      hr   =    60 min;
   end units;

subtype DELAY_LENGTH is TIME range 0 to TIME'HIGH;
impure function NOW return DELAY_LENGTH;

subtype NATURAL is INTEGER range 0 to INTEGER'HIGH;
subtype POSITIVE is INTEGER range 1 to INTEGER'HIGH;
type STRING is array (POSITIVE range <>) of CHARACTER;
type BIT_VECTOR is array (NATURAL range <>) of BIT;
type FILE_OPEN_KIND is (
   READ_MODE, WRITE_MODE, APPEND_MODE);
type FILE_OPEN_STATUS is (
   OPEN_OK, STATUS_ERROR, NAME_ERROR, MODE_ERROR);
attribute FOREIGN: STRING;

end STANDARD;
```

TEXTIO PACKAGE

The package **TEXTIO** contains declarations of type and subprograms that support formatting operations on text files. This package is shown here because it is a standard package in VHDL. Its declarations cannot be used in a synthesizeable design.

```
package TEXTIO is

type LINE is access STRING;
type TEXT is file of STRING;
type SIDE is (RIGHT, LEFT);
subtype WIDTH is NATURAL;

file INPUT: TEXT open READ_MODE is "STD_INPUT";
file OUTPUT: TEXT open WRITE_MODE is "STD_OUTPUT";
```

procedure READLINE (file F:TEXT; L: out LINE);

procedure READ (L:inout LINE;VALUE: out BIT; GOOD: out BOOLEAN);
procedure READ (L:inout LINE;VALUE: out BIT);

procedure READ (L:inout LINE;VALUE: out BIT_VECTOR; GOOD: out BOOLEAN);
procedure READ (L:inout LINE;VALUE: out BIT_VECTOR);

procedure READ (L:inout LINE;VALUE: out BOOLEAN; GOOD: out BOOLEAN);
procedure READ (L:inout LINE;VALUE: out BOOLEAN);

procedure READ (L:inout LINE;VALUE: out CHARACTER; GOOD: out BOOLEAN);
procedure READ (L:inout LINE;VALUE: out CHARACTER);

procedure READ (L:inout LINE;VALUE: out INTEGER; GOOD: out BOOLEAN);
procedure READ (L:inout LINE;VALUE: out INTEGER);

procedure READ (L:inout LINE;VALUE: out REAL; GOOD: out BOOLEAN);
procedure READ (L:inout LINE;VALUE: out REAL);

procedure READ (L:inout LINE;VALUE: out STRING; GOOD: out BOOLEAN);
procedure READ (L:inout LINE;VALUE: out STRING);

procedure READ (L:inout LINE;VALUE: out TIME; GOOD: out BOOLEAN);
procedure READ (L:inout LINE;VALUE: out TIME);

procedure WRITELINE (file F:TEXT; L:inout LINE);

procedure WRITE (L: inout LINE;VALUE: in BIT;
                JUSTIFIED: in SIDE:=RIGHT; FIELD: in WIDTH:=0);

procedure WRITE (L: inout LINE;VALUE: in BIT_VECTOR;
                JUSTIFIED: in SIDE:=RIGHT; FIELD: in WIDTH:=0);

procedure WRITE (L: inout LINE;VALUE: in BOOLEAN;
                JUSTIFIED: in SIDE:=RIGHT; FIELD: in WIDTH:=0);

procedure WRITE (L: inout LINE;VALUE: in CHARACTER;
                JUSTIFIED: in SIDE:=RIGHT; FIELD: in WIDTH:=0);

procedure WRITE (L: inout LINE;VALUE: in INTEGER;
                JUSTIFIED: in SIDE:=RIGHT; FIELD: in WIDTH:=0);

procedure WRITE (L: inout LINE;VALUE: in REAL;
                JUSTIFIED: in SIDE:=RIGHT; FIELD: in WIDTH:=0
                DIGITS: in NATURAL:= 0);

procedure WRITE (L: inout LINE;VALUE: in STRING;
                JUSTIFIED: in SIDE:=RIGHT; FIELD: in WIDTH:=0);

procedure WRITE (L: inout LINE;VALUE: in TIME;
                JUSTIFIED: in SIDE:=RIGHT; FIELD: in WIDTH:=0
                UNIT: in TIME:=ns);
end TEXTIO;

STD_LOGIC_1164 LIBRARY

```
--
-- sccsid @(#)std_logic_1164.vhdl 1.5 daisy(C) 10/14/92
--
-------------------------------------------------------------------------------------
--                                                                        --
--              PORTABLE - STD_LOGIC_1164                                  --
--                                                                        --
-- This package defines the portable constructs that were defined         --
-- by IEEE VHDL Model Standards Group.                                    --
--                                                                        --
-------------------------------------------------------------------------------------


-- ----------------------------------------------------------------
--
-- Title    : std_logic_1164 multivalue logic system
-- Library  : This package shall be compiled into a library
--          : symbolically named IEEE.
--          :
--          :
-- Developers: IEEE model standards group (par 1164)
-- Purpose  : This packages defines a standard for designers
--          : to use in describing the interconnection data types
--          : used in vhdl modelling.
--          :
-- Limitation: The logic system defined in this package may
--          : be insufficient for modelling switched transistors,
--          : since such a requirement is out of the scope of this
--          : effort. Furthermore, mathematics, primitives,
--          : timing standards, etc. are considered orthogonal
--          : issues as it relates to this package and are therefore
--          : beyond the scope of this effort.
--          :
-- Note     : No declarations or definitions shall be included in,
--          : or excluded from this package. The "package declaration"
--          : defines the types, subtypes and declarations of
--          : std_logic_1164. The std_logic_1164 package body shall be
--          : considered the formal definition of the semantics of
--          : this package. Tool developers may choose to implement
--          : the package body in the most efficient manner available
--          : to them.
--          :
-- ----------------------------------------------------------------
-- modification history :
-- ----------------------------------------------------------------
```

```
-- version | mod. date:|
--  v4.200 | 01/02/92 |
-- -------------------------------------------------------------------
library IEEE;
PACKAGE Std_logic_1164 is
-- Built-in attribute for synthesis:          --
attribute BUILT_IN: BOOLEAN;                  --
-----------------------------------------------------------------
-- Logic State System (unresolved)
-----------------------------------------------------------------
TYPE std_ulogic is ( 'U', -- Uninitialized
                     'X', -- Forcing Unknown
                     '0', -- Forcing 0
                     '1', -- Forcing 1
                     'Z', -- High Impedance
                     'W', -- Weak Unknown
                     'L', -- Weak 0
                     'H', -- Weak 1
                     '-'-- don't care
          );
attribute unique : string;
attribute unique of std_logic_1164 : package is "LOGIC_1164";
attribute unique of std_ulogic : type is "LOGIC9_BASE";


-----------------------------------------------------------------
-- These lines are added for synthesis: J. Bhasker, Feb 27, '92:
-----------------------------------------------------------------
ATTRIBUTE enum_type_encoding: STRING;
ATTRIBUTE enum_type_encoding OF std_ulogic: TYPE IS "U D 0 1 Z D 0
                                                     1 D";
-----------------------------------------------------------------
-- Unconstrained array of std_ulogic for use with the resolution function
-----------------------------------------------------------------
TYPE std_ulogic_vector IS ARRAY ( NATURAL RANGE <> ) of
std_ulogic;

attribute unique of std_ulogic_vector : type is "LOGIC9_BASE_VEC";


-----------------------------------------------------------------
-- Resolution function
-----------------------------------------------------------------
FUNCTION resolved ( s : std_ulogic_vector ) RETURN std_ulogic;
```

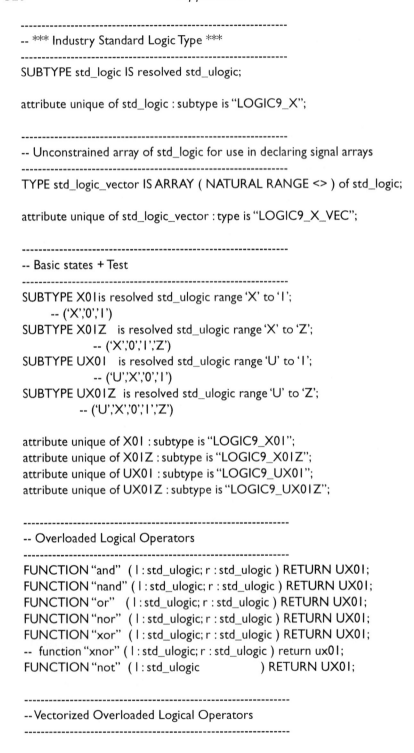

```
---------------------------------------------------------------
-- *** Industry Standard Logic Type ***
---------------------------------------------------------------
SUBTYPE std_logic IS resolved std_ulogic;

attribute unique of std_logic : subtype is "LOGIC9_X";


---------------------------------------------------------------
-- Unconstrained array of std_logic for use in declaring signal arrays
---------------------------------------------------------------
TYPE std_logic_vector IS ARRAY ( NATURAL RANGE <> ) of std_logic;

attribute unique of std_logic_vector : type is "LOGIC9_X_VEC";


---------------------------------------------------------------
-- Basic states + Test
---------------------------------------------------------------
SUBTYPE X01 is resolved std_ulogic range 'X' to '1';
      -- ('X','0','1')
SUBTYPE X01Z   is resolved std_ulogic range 'X' to 'Z';
             -- ('X','0','1','Z')
SUBTYPE UX01    is resolved std_ulogic range 'U' to '1';
             -- ('U','X','0','1')
SUBTYPE UX01Z  is resolved std_ulogic range 'U' to 'Z';
          -- ('U','X','0','1','Z')

attribute unique of X01 : subtype is "LOGIC9_X01";
attribute unique of X01Z : subtype is "LOGIC9_X01Z";
attribute unique of UX01 : subtype is "LOGIC9_UX01";
attribute unique of UX01Z : subtype is "LOGIC9_UX01Z";


---------------------------------------------------------------
-- Overloaded Logical Operators
---------------------------------------------------------------
FUNCTION "and"  ( l : std_ulogic; r : std_ulogic ) RETURN UX01;
FUNCTION "nand" ( l : std_ulogic; r : std_ulogic ) RETURN UX01;
FUNCTION "or"   ( l : std_ulogic; r : std_ulogic ) RETURN UX01;
FUNCTION "nor"  ( l : std_ulogic; r : std_ulogic ) RETURN UX01;
FUNCTION "xor"  ( l : std_ulogic; r : std_ulogic ) RETURN UX01;
-- function "xnor" ( l : std_ulogic; r : std_ulogic ) return ux01;
FUNCTION "not"  ( l : std_ulogic            ) RETURN UX01;


---------------------------------------------------------------
-- Vectorized Overloaded Logical Operators
---------------------------------------------------------------
```

FUNCTION "and" (l, r : std_logic_vector) RETURN std_logic_vector;
FUNCTION "nand" (l, r : std_logic_vector) RETURN std_logic_vector;
FUNCTION "or" (l, r : std_logic_vector) RETURN std_logic_vector;
FUNCTION "nor" (l, r : std_logic_vector) RETURN std_logic_vector;
FUNCTION "xor" (l, r : std_logic_vector) RETURN std_logic_vector;
FUNCTION "not" (l: std_logic_vector) RETURN std_logic_vector;

FUNCTION "and" (l, r : std_ulogic_vector) RETURN
std_ulogic_vector;
FUNCTION "nand" (l, r : std_ulogic_vector) RETURN
std_ulogic_vector;
FUNCTION "or" (l, r : std_ulogic_vector) RETURN
std_ulogic_vector;
FUNCTION "nor" (l, r : std_ulogic_vector) RETURN
std_ulogic_vector;
FUNCTION "xor" (l, r : std_ulogic_vector) RETURN
std_ulogic_vector;
FUNCTION "not" (l : std_ulogic_vector) RETURN
std_ulogic_vector;

```
-- ---------------------------------------------------------------------
-- Note : The declaration and implementation of the "xnor" function is
-- specifically commented until at which time the VHDL language has
been
-- officially adopted as containing such a function. At such a point,
-- the following comments may be removed along with this notice
without
-- further "official" ballotting of this std_logic_1164 package. It is
-- the intent of this effort to provide such a function once it becomes
-- available in the VHDL standard.
-- ---------------------------------------------------------------------
-- function "xnor" ( l, r : std_logic_vector ) return std_logic_vector;
-- function "xnor" ( l, r : std_ulogic_vector ) return std_ulogic_vector;
-----------------------------------------------------------------
-- Conversion Functions
-----------------------------------------------------------------
```
FUNCTION To_bit (s : std_ulogic;   xmap : BIT := '0')
        RETURN BIT;
FUNCTION To_bitvector (s : std_logic_vector ; xmap : BIT := '0')
        RETURN BIT_VECTOR;
FUNCTION To_bitvector (s : std_ulogic_vector; xmap : BIT := '0')
        RETURN BIT_VECTOR;

FUNCTION To_StdULogic   (b : BIT   )
        RETURN std_ulogic;

```
FUNCTION To_StdLogicVector  ( b : BIT_VECTOR      )
       RETURN std_logic_vector;
FUNCTION To_StdLogicVector  ( s : std_ulogic_vector )
       RETURN std_logic_vector;
FUNCTION To_StdULogicVector ( b : BIT_VECTOR      )
       RETURN std_ulogic_vector;
FUNCTION To_StdULogicVector ( s : std_logic_vector  )
       RETURN std_ulogic_vector;


-----------------------------------------------------------------
-- strength strippers and type convertors
-----------------------------------------------------------------
FUNCTION To_X0I  ( s : std_logic_vector ) RETURN std_logic_vector;
FUNCTION To_X0I  ( s : std_ulogic_vector) RETURN
std_ulogic_vector;
FUNCTION To_X0I  ( s : std_ulogic  ) RETURN X0I;
FUNCTION To_X0I  ( b : bit_vector  ) RETURN std_logic_vector;
FUNCTION To_X0I  ( b : bit_vector  ) RETURN std_ulogic_vector;
FUNCTION To_X0I  ( b : bit ) RETURN X0I;

FUNCTION To_X0IZ ( s : std_logic_vector ) RETURN std_logic_vector;
FUNCTION To_X0IZ ( s : std_ulogic_vector) RETURN
std_ulogic_vector;
FUNCTION To_X0IZ ( s : std_ulogic  ) RETURN X0IZ;
FUNCTION To_X0IZ ( b : bit_vector  ) RETURN std_logic_vector;
FUNCTION To_X0IZ ( b : bit_vector  ) RETURN std_ulogic_vector;
FUNCTION To_X0IZ ( b : bit     ) RETURN X0IZ;
FUNCTION To_UX0I ( s : std_logic_vector ) RETURN std_logic_vector;
FUNCTION To_UX0I ( s : std_ulogic_vector) RETURN
std_ulogic_vector;
FUNCTION To_UX0I ( s : std_ulogic  ) RETURN UX0I;
FUNCTION To_UX0I ( b : bit_vector  ) RETURN std_logic_vector;
FUNCTION To_UX0I ( b : bit_vector  ) RETURN std_ulogic_vector;
FUNCTION To_UX0I ( b : bit ) RETURN UX0I;
attribute BUILT_IN of TO_BIT: function is TRUE;
attribute BUILT_IN of TO_BITVECTOR: function is TRUE;
attribute BUILT_IN of TO_STDULOGIC: function is TRUE;
attribute BUILT_IN of TO_STDULOGICVECTOR: function is TRUE;
attribute BUILT_IN of TO_STDLOGICVECTOR: function is TRUE;
attribute BUILT_IN of TO_X0I: function is TRUE;
attribute BUILT_IN of TO_X0IZ: function is TRUE;
attribute BUILT_IN of TO_UX0I: function is TRUE;
```

```
-----------------------------------------------------------------
-- Edge Detection
-----------------------------------------------------------------
FUNCTION rising_edge  (SIGNAL s : std_ulogic) RETURN boolean;
FUNCTION falling_edge (SIGNAL s : std_ulogic) RETURN boolean;

-- synthesis built-in functions
attribute BUILT_IN of rising_edge : function is TRUE;
attribute BUILT_IN of falling_edge : function is TRUE;
-----------------------------------------------------------------
-- object contains an unknown
-----------------------------------------------------------------
FUNCTION Is_X ( s : std_ulogic_vector ) RETURN  BOOLEAN;
FUNCTION Is_X ( s : std_logic_vector  ) RETURN  BOOLEAN;
FUNCTION Is_X ( s : std_ulogic) RETURN  BOOLEAN;

END Std_logic_1164;
```

Index

Underlined references are to figures; _underlined italics_ refers to boxes; **_un-derlined italic bold_** refers to tables.